HAMLYN'S ALL-COLOUR BOOK OF

Freezer Cookery

HAMLYN
LONDON · NEW YORK · SYDNEY · TORONTO

Contents

Useful facts and figures	4
All about freezing	6
Freezing vegetables	8
Freezing fruit	9
Basic freezing recipes	10
Soups and starters	17
Fish, meat and poultry	24
Vegetable dishes	36
Desserts and gâteaux	39
Party entertaining	50
Picnics and packed lunches	54
Cakes, biscuits and bread	57
Index	63

Acknowledgements
Recipes created by Moya Maynard
Photography by John Lee
Cover picture by Iain Reid
Artwork by John Scott Martin
China kindly loaned by Royal Doulton Tableware Limited
(Lambeth Stoneware)
Foil products by Alcan Foil

Published by
The Hamlyn Publishing Group Limited
LONDON · NEW YORK · SYDNEY · TORONTO
Astronaut House, Feltham, Middlesex, England

© Copyright The Hamlyn Publishing Group Limited 1977
ISBN 0 600 33611 5

Printed in England by Jarrold and Sons Limited, Norwich

Useful facts and figures

Notes on metrication

In this book quantities are given in metric, imperial and American measures. Exact conversion from imperial to metric measures does not usually give very convenient working quantities and so the metric measures have been rounded off into units of 25 grams. The table below shows the recommended equivalents.

Ounces	Approx g to nearest whole figure	Recommended conversion to nearest unit of 25
1	28	25
2	57	50
3	85	75
4	113	100
5	142	150
6	170	175
7	198	200
8	227	225
9	255	250
10	283	275
11	312	300
12	340	350
13	368	375
14	397	400
15	425	425
16 (1 lb)	454	450
17	482	475
18	510	500
19	539	550
20	567	575

Note: When converting quantities over 20 oz first add the appropriate figures in the centre column, then adjust to the nearest unit of 25. As a general guide, 1 kg (1000 g) equals 2.2 lb or about 2 lb 3 oz. This method of conversion gives good results in nearly all cases but in certain baking recipes a more accurate conversion is necessary to produce a balanced recipe. On the other hand, quantities of such ingredients as vegetables, fruit, meat and fish which are not critical are rounded off to the nearest quarter of a kg as this is how they are likely to be purchased.

Liquid measures The millilitre has been used in this book and the following table gives a few examples.

Imperial	Approx ml to nearest whole figure	Recommended ml
$\frac{1}{4}$ pint	142	150 ml
$\frac{1}{2}$ pint	283	300 ml
$\frac{3}{4}$ pint	425	450 ml
1 pint	567	600 ml
1$\frac{1}{2}$ pints	851	900 ml
1$\frac{3}{4}$ pints	992	1000 ml (1 litre)

Note: For quantities of 1$\frac{3}{4}$ pints and over we have used litres and fractions of a litre.

Spoon measures All spoon measures given in this book are level.

Can sizes At present, cans are marked with the exact (usually to the nearest whole number) metric equivalent of the imperial weight of the contents, so we have followed this practice when giving can sizes.

Oven temperatures

The table below gives recommended equivalents.

	°F	°C	Gas Mark
Very cool	225	110	$\frac{1}{4}$
	250	120	$\frac{1}{2}$
Cool	275	140	1
	300	150	2
Moderate	325	160	3
	350	180	4
Moderately hot	375	190	5
	400	200	6
Hot	425	220	7
	450	230	8
Very hot	475	240	9

Note: When making any of the recipes in this book, only follow one set of measures as they are not interchangeable.

Notes for American users

Although the recipes in this book give American measures, the lists below give some equivalents or substitutes for terms and commodities which may be unfamiliar to American readers.

Equipment and terms
BRITISH/AMERICAN
cake or loaf tin/cake or loaf pan
cling film/saran wrap
cocktail stick/toothpick
flan tin/pie pan
frying pan/skillet
greaseproof paper/waxed paper
grill/broil
kitchen paper/paper towels
liquidise(r)/blend(er)
mince/grind
packet/package
piping bag/pastry bag
polythene/plastic
roasting tin/roasting pan
sandwich tin/layer cake pan
star nozzle/fluted nozzle
stoned/pitted
tartlet tins/patty shells
top and tail/stem and head

Ingredients
BRITISH/AMERICAN
Arctic Roll/ice cream cake roll
aubergine/eggplant
beetroot/beet
belly of pork/salt pork
bicarbonate of soda/baking soda
biscuit/cookie or cracker
black treacle/molasses
cake mixture/cake batter
castor or granulated sugar/sugar
celery stick/celery stalk
cocoa powder/unsweetened cocoa
cooking apple or pear/baking apple or pear
cornflour/cornstarch
courgette/zucchini
crystallised ginger/candied ginger
demerara sugar/brown sugar
double cream/heavy cream
dried milk powder/dried milk solids
essence/extract
gelatine/gelatin
gherkin/sweet dill pickle
gingernut/gingersnap
glacé cherry/candied cherry
golden syrup/maple syrup
ham/cured or smoked ham
head celery/bunch celery
icing/frosting
icing sugar/confectioners' sugar
ketchup/catsup
lard/shortening
marrow/squash
mixed peel/candied peel
mould/mold
natural yogurt/unflavored yogurt
orange jelly/orange-flavored gelatin
peeled prawn/shelled shrimp
pig's liver/pork liver
pimento/pimiento
plain chocolate/semi-sweet chocolate
plain flour/all-purpose flour
pork fillet/pork tenderloin
puff pastry/puff paste
scampi/jumbo shrimp
scone/biscuit
self-raising flour/self-rising flour, or all-purpose flour sifted with baking powder
shortcrust pastry/basic pie dough
single cream/light cream
sorbet/sherbet
soured cream/sour cream
spring onion/scallion
stem ginger/preserved ginger
streaky bacon rasher/bacon slice
sultana/seedless white raisin
Swiss roll/jelly roll
tomato purée/tomato paste
unsalted butter/sweet butter

Notes for Australian users

Ingredients in this book are given in cup, metric and imperial measures. In Australia the American 8-oz measuring cup is used in conjunction with the imperial pint of 20 fluid ounces. It is most important to remember that the Australian tablespoon differs from both the British and American tablespoons; the table below gives a comparison between the standard tablespoons used in the three countries. The British standard tablespoon holds 17.7 millilitres, the American 14.2 millilitres, and the Australian 20 millilitres. A teaspoon holds approximately 5 millilitres in all three countries.

British	American	Australian
1 teaspoon	1 teaspoon	1 teaspoon
1 tablespoon	1 tablespoon	1 tablespoon
2 tablespoons	3 tablespoons	2 tablespoons
3$\frac{1}{2}$ tablespoons	4 tablespoons	3 tablespoons
4 tablespoons	5 tablespoons	3$\frac{1}{2}$ tablespoons

Note: The British and Australian pint is 20 fluid ounces as opposed to the American pint which is 16 fluid ounces.

Freezing storage times

Cooked dishes

Soups, sauces and stocks	4 months
Highly seasoned sauces	2 months
Fish dishes	2 months
Meat loaves and pâtés	1 month
Meat casseroles and curries	2 months
Pastry dishes	2 months
Vegetable dishes	2 months
Cakes, scones and biscuits	6 months
Bread	1 month

Uncooked dishes

Asparagus and smoked salmon rolls, pâté pinwheels	2 months
Beefburgers with onion	2 months
Shoulder of lamb	9–12 months
Sorbets and ice creams	3 months
Pineapple pudding mixture	1 month

All about freezing

When you are thinking of buying a freezer the first things to decide are where you will put it and which type you would prefer. Any well-ventilated, dry place with a strong floor is suitable, as long as there is an electric socket nearby. Floor area is usually the main consideration when choosing between a chest and an upright but there are several other factors to be borne in mind. A chest freezer offers more storage space for less money, is more economical to run because less cold air escapes when the lid is opened, and only needs to be defrosted once a year. An upright is more convenient, easier to load and unload and more likely to fit into the average kitchen, but it will need to be defrosted twice a year unless you buy a model which defrosts automatically.

Packaging materials

Once you have purchased and installed your freezer, and are faced with the exciting task of stocking it, one of the most important rules to remember is that all food stored in the freezer must be packed properly in moisture/vapour-proof material, and tightly sealed. This is absolutely essential to prevent it losing moisture and becoming dry. Careless packing can result in freezer burn appearing particularly on meat, fish and poultry. This causes toughness and loss of flavour and produces greyish-white marks on the surface. Good packaging also stops cross-flavouring between strongly flavoured foods.

There are many different packaging materials to choose from:

Foil Heavy-duty freezer foil is the best, but ordinary foil can be used double. It is easy to handle and makes an effective wrapping material for unevenly shaped foods like meat, poultry and fish. Be sure to mould it closely to exclude all air, and seal the edges with freezer tape.

Foil can also be used for lining casseroles so that the contents can be removed when frozen and the dish used again. Line the dish carefully with foil, leaving a large 'frill' of extra foil around the top. Pour in the food. This can then be cooked in the lined dish or it may be more convenient to cook a large quantity in one big casserole and then divide it between several smaller foil-lined dishes for freezing in meal-size portions. Cool and freeze until solid. Lift the foil-wrapped block out of the dish, fold the foil 'frill' over the top and seal with freezer tape. Overwrap with a polythene bag before returning to the freezer. To serve, strip off the foil and place the food in the casserole again for thawing and reheating.

Alternatively, cooked food can be frozen in the dish in which it was cooked. When the food is solid, dip the dish quickly into hot water, turn out the frozen block and wrap it in foil.

Freezerproof china Ovenproof china which can withstand long-term storage in the freezer is now available. This means that, after cooling, covering with foil, sealing and labelling, food can be transferred to the freezer in the dish in which it was cooked.

Freezer tape Special adhesive tape which does not peel off at low temperatures. Particularly useful for sealing foil packages and polythene bags. The type which looks like masking tape has the advantage that it can be written on but the clear variety can be used to protect paper labels.

Foil bags Particularly suitable for storing liquid and semi-liquid foods like soups or stews.

Foil dishes These are very useful for freezing cooked food as it can be reheated in the same dish. They are available in many different shapes and sizes, some of which have lids. The dishes can be re-used after careful washing and drying. If the lids are not re-usable the tops can be covered with double-thickness foil and then sealed.

Polythene bags Heavy-duty extra thick polythene bags should be used for freezing – the coloured ones are particularly useful for easy identification. Polythene bags are very popular for packing solids such as vegetables, pastries, cakes, meat, fish and poultry, and as added protection around foil-wrapped parcels. As much air as possible should be extracted and the bag sealed with a twist tie or freezer tape. They can be washed and used again but are best as overbags after the first time.

If they are to be used for liquids, it is a good idea to form them into a square shape to save storage space. Line a square polythene container with a polythene bag. Pour in the food to be frozen and seal, leaving a headspace. Freeze until hard, then remove the wrapped frozen food from the container.

Rigid polythene containers These can be re-used indefinitely but do make sure that they are guaranteed to withstand the low temperatures without warping or cracking. They come in a variety of shapes and sizes but the square ones take up the least freezer space. They are ideal for storing casseroles, soups or other liquid foods, but be sure to leave a space to allow for expansion. Delicate items like decorated cakes are well protected in the larger polythene containers.

Boil-in-bags These are made of specially thick polythene in which food can be frozen and then cooked or reheated in boiling water. They must be heat-sealed with a heat sealer or iron.

Cling film Freezer cling film is much thicker and stronger than the ordinary film, and is very good for moulding around difficult shapes. Ordinary cling film is useful for separating individual items like chops or steaks before they are over-wrapped.

Twist ties These are made of wire covered with either paper or plastic and are used for sealing the tops of polythene bags or opened bags of commercially frozen foods.

Labels Some are adhesive and come in various sizes and colours. Be sure to stick them on to polythene bags before filling. Labels attached to twist ties are useful for shapes on which it is difficult to write or stick labels.

Heat sealers These are useful if you freeze a lot of produce in heavy-gauge polythene bags. They seal the two pieces of polythene together, providing an airtight seal.

Freezing tips

Labelling and recording You may think that you will remember what you have put in the freezer, and what you have already used, but you will soon find that this is extremely difficult. It is much better to label everything clearly with the name of the food, the date it was frozen, the quantity or the number it is to serve. A log is also a very helpful guide as to when restocking is necessary.

Fast freezing It is very important to freeze down food as quickly as possible because slow freezing damages the cell structure which may result in a loss of nutrients, flavour and texture when the food is thawed or cooked. The fast freeze switch by-passes the thermostat, causing continuous running to bring the temperature below the normal storage temperature of −18°C/0°F. It should be switched on at least 2 hours before packing fresh food into the freezer, and left on for a further 24 hours.

It is not worth using the fast freeze switch if you are only freezing down a very small weight of food, say 1 kg/2 lb or less. When loading fresh food for freezing, ensure that it is not placed in contact with any already frozen food and that, as far as possible, it is put on refrigerated shelves in an upright freezer or against the walls of a chest freezer.

Only freeze within 24 hours the weight of food recommended by the freezer manufacturer and if the maximum weight is to be frozen, operate the fast freeze switch according to the manufacturer's recommendations. If weights are not given, do not exceed 10 per cent of the loading capacity. It is generally accepted that an upright freezer holds 7.5 kg/15 lb per 28.3 litres/1 cubic foot and a chest holds 10 kg/20 lb per 28.3 litres/1 cubic foot so this figure should be multiplied by the net capacity of the freezer to find out the weight of the food which can be frozen down at one time.

Open freezing Commercially frozen fruit and vegetables are 'blast' frozen to keep them separate and free flowing. To keep them separate at home, they have to be spread out on trays and frozen until hard before being packed. It is also a good idea to open freeze very delicate items like decorated cakes and soufflés before packing in rigid polythene containers. Fragile foods should not be packed in bags as they can easily be damaged even when frozen.

Dividers Another way of preventing food from freezing together is to place a sheet of foil, cling film, waxed paper or greaseproof paper between such items as chops, steaks, sausages, pastry rounds or pancakes before packing them in polythene bags. This way the required number can be removed easily when needed.

Freezing cooked dishes One of the greatest joys of owning a freezer is the way that it enables you to cook in bulk when it is convenient and you have the time, or to prepare for a dinner party in advance and save all that last-minute panic.

Instructions on freezing and thawing have been given with the individual recipes in this book but it must also always be remembered that all cooked food must be completely cooled as quickly as possible and then frozen.

Food which has already been frozen can be used with complete safety in dishes which are to be frozen as long as it is thoroughly cooked in between and not left lying around in a warm kitchen.

Rigid polythene or foil containers are the most popular packaging materials for freezing cooked dishes but sheet foil can also be used, as already explained.

Excluding air All air must be expelled from packs containing solid food. Foil and cling film should be moulded closely and air must be withdrawn from bags with the aid of a pump or by dipping the bag into water to expel the air – dry before freezing. Fill spaces above solids packed in rigid containers with a piece of crumpled foil.

Headspace Remember that liquids expand on freezing so always leave a headspace of 1–2.5 cm/½–1 inch above casseroles, soups, sauces or any other liquids.

Defrosting This operation is made easier if it is undertaken when freezer stocks are low, but it is not essential to run the food down deliberately.

Disconnect the freezer and remove all the food. Small items and those which thaw quickly (e.g. ice cream) can be kept in a refrigerator. The remainder will keep perfectly well for the duration of the defrosting if they are packed closely together and covered with plenty of newspaper and a thick blanket.

Place a bowl of hot water in the bottom of the freezer to loosen the ice, then remove the ice from the sides and shelves with a plastic scraper. Do not use metal implements because they damage the interior surface. The ice can either be collected up in a thick towel placed in the bottom of the freezer, or with a dustpan and brush.

Wipe out with a solution of 1 tablespoon bicarbonate of soda to each generous litre/2 pints/2½ pints water, rinse, and dry with a clean absorbent cloth. Reconnect and replace the food. The whole defrosting process should not take more than 2 hours.

To keep the outside of the cabinet in good condition and to prevent it from rusting it is a good idea to give it an occasional coat of wax polish.

Emergencies and power cuts If you tape over the switch and plug and make sure that the freezer is never switched off, emergencies are extremely unlikely. Most freezers are very reliable and as a full load of food will remain in perfect condition in the cabinet for over 12 hours after the power has been turned off (provided the lid or door is not opened), there is little need to worry about power cuts. However, it is wise to take out a freezer insurance and to have a good maintenance engineer on call.

When moving house it is sensible to reduce the stocks and to make sure that the freezer is the last thing to go into the removal van and the first to come out at your destination. If the journey is to take more than 12 hours the freezer should be emptied completely if the removal firm cannot arrange for it to be connected to an electrical supply overnight.

Freezing vegetables

Most vegetables freeze well and keep for up to a year provided they are blanched. Vegetables begin to deteriorate as soon as they are gathered and would continue to do so, even when frozen, if they were not blanched. Blanching halts the spoilage caused by enzymes and preserves the colour and flavour of the food. There are a few vegetables which can be frozen for a short time unblanched but they must not be stored for longer than is indicated in the following chart. Vegetables which are to be frozen cooked do not need to be blanched.

Blanching

Place 0.5 kg/1 lb/1 lb prepared vegetables in a blanching basket and plunge into a large saucepan of boiling water. Time the blanching from the moment the water returns to a full boil. This must be done accurately as under-blanching will not completely destroy enzyme action and over-blanching can make the vegetables soft and mushy. Remove the basket and cool the vegetables immediately under cold running water or in a bowl of iced water for the same length of time as they have been blanched. Drain thoroughly and dry on absorbent paper. Pack in polythene bags or containers in meal-sized portions. Open freeze first if free-flowing vegetables are required.

Preparation	Blanching time
Artichoke, globe Remove stalk and outer leaves. Wash. Blanch whole. Drain upside down. Pack individually.	7 minutes
Artichoke, Jerusalem Freeze as cooked purée for soups.	–
Asparagus Cut into lengths to fit chosen container. Wash, grade, blanch and tie in bundles. Pack bundles tips to stems.	Thin spears 2 minutes Thick spears 4 minutes
Aubergines Peel, cut into 2.5-cm/1-inch slices. Blanch, cool, open freeze then pack with waxed paper between layers.	3 minutes
Avocados Freeze as uncooked purée, adding seasoning and lemon juice. Keep for 3 months only. Use for soups or party dips.	–
Beans, broad Pod and blanch.	3 minutes
Beans, French Top, tail and wash. Slice thickly, or leave whole if small. Blanch. (Small French beans can be frozen whole unblanched for up to 3 months.)	Sliced 2 minutes Whole 3 minutes
Beans, runner String, top, tail and wash. Slice thickly and blanch.	2 minutes
Beetroot Choose small beetroots. Cook in boiling, salted water until tender. Peel and pack whole if very small, slice if slightly larger.	–
Broccoli Remove outer leaves and woody stems. Cut into sprigs, wash and blanch.	Thin stems 2 minutes Thick stems 4 minutes
Brussels sprouts Choose firm, small sprouts. Grade into sizes. Trim off outer leaves and woody stems. Wash and blanch.	Small 3 minutes Medium 4 minutes
Cabbage Choose crisp, firm cabbage. Shred, wash and blanch. Drain well.	1½ minutes
Carrots Trim, scrape and wash. Freeze small, young carrots whole; slice or dice larger ones. Blanch. (Young tender carrots can be frozen whole unblanched for up to 3 months.)	3 minutes
Cauliflower Choose tight, white, unblemished heads. Break into small florets. Wash and blanch, adding lemon juice to the blanching water to keep the cauliflower a good colour.	3 minutes
Celery Trim, clean and cut into 1-cm/½-inch lengths. Blanch. Use only in cooked dishes.	3 minutes
Corn on the cob Choose young fresh cobs. Remove husk and silk. Cut off stem. Blanch.	Small cobs 5 minutes Large cobs 8 minutes
Courgettes Halve if small; cut into 1-cm/½-inch slices if larger. Blanch. (Can be frozen whole unblanched for up to 3 months.)	1 minute
Herbs Useful for flavouring casseroles, soups and sauces. Chop, cover with water in ice cube trays, freeze then pack the cubes. Or freeze in sprigs which can be crumbled while still frozen to save chopping. For mint sauce, pour sugar syrup (see page 9) instead of water over chopped mint in ice cube trays. When required, thaw and mix with a little vinegar.	–
Leeks Remove tops, roots and coarse outer leaves. Cut into 1-cm/½-inch slices and wash well. Sauté in butter for 3–4 minutes or blanch. (Can be frozen unblanched for up to 3 months.)	1 minute
Marrow Choose young marrows. Peel, cut into 2.5-cm/1-inch slices and remove seeds. Blanch.	2 minutes
Mushrooms Wipe clean. Leave whole if button; slice if larger. Sauté in butter for 1 minute. (Can be frozen raw for up to 3 months.)	–
Onions Peel. Leave whole if button; chop if larger. Blanch. (Can be frozen chopped unblanched for up to 3 months.)	Chopped 2 minutes Whole 6 minutes
Parsnips Trim and peel. Leave whole if very small; slice or dice if larger. Blanch.	Sliced 2 minutes Whole 3 minutes
Peas Shell and blanch.	1½ minutes
Peppers Wash. Remove stem and seeds and chop or slice flesh. Blanch. (Can be frozen whole or chopped unblanched for up to 3 months.)	3 minutes
Potatoes Freeze cooked – boiled, roasted, jacket, duchesse or croquette – preparing in the usual way. Or freeze as chips after blanching in oil and draining on absorbent paper.	2 minutes
Spinach Wash well and blanch in 100-g/4-oz/¼-lb batches. Drain very well and chop if liked.	2 minutes
Tomatoes Can be frozen whole for use in cooked dishes, removing skins before beginning to thaw. Better if skinned, simmered and frozen as purée.	–
Turnips Trim and peel. Leave whole if very small; slice or dice if larger. Blanch.	Sliced 2 minutes Whole 3 minutes

NOTE: All vegetables (except corn on the cob) can be cooked from frozen but be sure to reduce the normal cooking time if they have been blanched.

Freezing fruit

Freezing is definitely the best and simplest way of preserving fruit. It is easy to prepare and has a long storage life – usually up to 1 year. However, for good results only the best quality, ripe, firm fruit should be used, unless it is to be frozen as purée.

Different fruits need to be prepared in different ways and it is a good idea to think about how you will eventually use the fruit as this will influence the way it is packed.

Methods of preparing fruit

Dry pack (with or without sugar) The fruit can be open frozen and then packed in polythene bags or rigid containers with no sugar added. This gives the most natural flavour when the fruit is thawed, and is suitable for soft fruits like blackberries, cranberries, loganberries, raspberries and strawberries.

Alternatively, the fruit can be mixed with about a quarter of its weight of sugar when being packed into polythene bags or containers. This method is suitable for currants, gooseberries and rhubarb but these fruits can also be frozen without sugar if they are to be used unsweetened.

Sugar syrup Used for fruits which discolour easily when exposed to the air, e.g. apricots, peaches and pears. Syrups vary in strength between 50 g/2 oz/¼ cup sugar to 600 ml/1 pint/2½ cups water and 0.5 kg/1 lb/2 cups sugar to 600 ml/1 pint/2½ cups water. The proportion you choose will depend on personal preference and the natural sweetness of the fruit. The most usual syrups have between 225 g/8 oz/1 cup and 350 g/12 oz/1½ cups sugar per 600 ml/1 pint/2½ cups water.

Dissolve the sugar in the water and boil for 1 minute. Chill before using. Add ¼ teaspoon ascorbic acid to each 600 ml/1 pint/2½ cups syrup to prevent discoloration on thawing. Alternatively, sprinkle the fruit with lemon juice. Half-fill rigid polythene containers with the syrup and add the prepared fruit, leaving a headspace. Top with a piece of crumpled greaseproof paper, foil or polythene to push down any exposed fruit and prevent discoloration.

Purée Over-ripe fruit like raspberries and strawberries can be sieved and frozen as uncooked purée for use in ice creams, soufflés or sauces.

Fruits like apples, apricots and gooseberries can be cooked with or without sugar in a little water, sieved or liquidised if necessary, and frozen as purée in rigid polythene containers, foil bags or polythene bags, for use in sauces or cooked desserts.

Preparation	Uses
Apples Peel, core and slice into salted cold water. Blanch for 1 minute, drain then pack with or without sugar. Or cook with sugar and a little water until soft. Or pack as unsweetened cooked purée.	Pies and puddings Pies and puddings Apple sauce, baby food and desserts
Apricots Wash, skin, halve and stone or leave whole. Pack in syrup with lemon juice or ascorbic acid. Or cook with sugar and a little water until soft. Or pack as cooked purée.	Flans and fruit salads Pies and flans Sauce for ice cream
Blackberries Wash only if necessary, drain and hull. Open freeze then pack.	As for fresh
Blackcurrants Stem and wash if necessary. Open freeze then pack, or dry pack with sugar.	Pies and puddings
Cherries Remove stalks and wash. Open freeze whole fruit then pack, or dry pack with sugar. Or pack in syrup.	Fruit salads Pies and puddings
Damsons Wash, halve and stone. Pack in syrup.	Pies and puddings
Gooseberries Top, tail and wash. Dry pack with sugar or pack in syrup. Or pack as cooked purée.	Pies and puddings Fools or other desserts
Grapefruit Peel, remove pith and segment or slice flesh. Dry pack with sugar or pack in syrup.	Grapefruit salads or cocktails
Grapes Wash, leave whole if seedless; skin, halve and pip if not. Pack in syrup.	Fruit salads
Greengages Wash, halve and stone. Pack in syrup.	Pies and puddings
Lemons Slice, open freeze then pack. Or freeze whole. Or grate rind and mix with sugar. Or freeze juice in ice cube trays then pack.	In drinks As for fresh for juice and grated rind Flavouring and decoration Flavouring
Loganberries Wash only if necessary, drain and hull. Open freeze then pack.	As for fresh

Preparation	Uses
Melons Cut flesh into balls or cubes. Dry pack with sugar or pack in syrup.	Fruit salads
Oranges Peel, remove pith and segment or slice flesh. Dry pack with sugar or pack in syrup. Or grate rind and mix with sugar. Or freeze juice in ice cube trays then pack. Freeze Seville oranges whole.	Fruit salads or puddings Flavouring and decoration Flavouring Marmalade
Peaches Peel, halve and stone. Pack in syrup with lemon juice or ascorbic acid.	Fruit salads and other desserts
Pears Peel and core, cut into quarters or slices. Poach in syrup for 1½–2 minutes. Pack in cooled syrup with lemon juice or ascorbic acid. (It is not worth freezing pears unless you have a very heavy crop.)	Fruit salads and other desserts
Pineapple Peel, core and cut into slices or cubes. Pack in its own juice or in syrup.	As for fresh
Plums Wash, halve and stone. Pack in syrup.	Pies and puddings
Raspberries Wash only if necessary, drain and hull. Open freeze then pack. Or purée by sieving if slightly soft.	As for fresh Ice creams, sorbets, soufflés, sauces
Rhubarb Trim, wash and cut into 2.5-cm/1-inch lengths. Dry pack with or without sugar. Or cook with sugar and a little water until tender.	Pies and puddings Pies and puddings
Strawberries Choose small fruit. Wash only if necessary, drain and hull. Open freeze then pack, or dry pack with sugar. Or purée if slightly soft.	Fruit salads and other desserts Ice creams, sorbets, soufflés, sauces

NOTE: Fruit to be served uncooked should be thawed overnight in the refrigerator or for 3–4 hours at room temperature. Soft fruits should still be slightly chilled when served to prevent them becoming mushy. Fruit to be cooked can be thawed first or heated gently from frozen.

Basic freezing recipes

It is an excellent idea to keep in the freezer a supply of basics such as a meat sauce which can be thawed out and added to other ingredients to make a variety of quick, interesting dishes. It is also always handy to have in stock a basic sandwich cake to serve with tea or coffee and a few savouries to accompany drinks.

Meat sauce

METRIC/IMPERIAL/AMERICAN
1 large onion, chopped
1 tablespoon corn oil
0.5 kg/1 lb/2 cups minced beef
40 g/1½ oz/6 tablespoons plain flour
100 g/4 oz/1 cup mushrooms, chopped
1–2 cloves garlic, crushed
1 (227-g/8-oz/8-oz) can peeled tomatoes
2 tablespoons/2 tablespoons/3 tablespoons tomato purée
1 bay leaf
150 ml/¼ pint/⅔ cup beef stock

Place the onion and oil in a saucepan and cook gently for about 10 minutes. Add the mince and cook until no longer red. Blend in the flour and the remaining ingredients.
  Cover and simmer very gently for 40 minutes.
To serve at once: Use half the above quantity of sauce to make the meat and vegetable pasties (see page 28) or pancake layer pie (see page 27). Use the full quantity for the individual shepherd's pies (see page 29).
To freeze: Put into rigid polythene containers. Seal, label and freeze.
To thaw: Leave at room temperature for 3–4 hours, in the refrigerator overnight, or slowly heat through from frozen in a saucepan.

Serves 4 as a sauce for spaghetti
Makes 4 pasties, or 1 meat and pancake pie to serve 6

All-in-one sandwich cake

METRIC/IMPERIAL/AMERICAN
110 g/4 oz/½ cup soft margarine
110 g/4 oz/½ cup castor sugar
2 large eggs
110 g/4 oz/1 cup self-raising flour
1 teaspoon baking powder
TO FINISH:
4 tablespoons/4 tablespoons/⅓ cup freezer jam
100 ml/4 fl oz/½ cup double cream, whipped (optional)
icing sugar for dusting
walnuts, glacé cherries, angelica

Line the bases of two 18-cm/7-inch sandwich tins with greaseproof paper. Grease, and sprinkle with flour.
Using a wooden spoon, beat all the ingredients together for 2–3 minutes. Divide the mixture between the tins.
Bake in a moderate oven (160°C, 325°F, Gas Mark 3) for 25–35 minutes until firm. Allow to stand for a few minutes before turning out on to a cake rack. Leave to become cold.
To serve at once: Sandwich the cakes together with freezer jam and whipped double cream, if liked. Dust with icing sugar and decorate with walnuts, glacé cherries and angelica.
To freeze: Interleave the layers with freezer layer tissue and wrap in foil. Seal, label and freeze.
To thaw: Leave in the foil at room temperature for 2 hours, then finish as above.

Makes 1 (18-cm/7-inch) round cake

Freezer jam

METRIC/IMPERIAL/AMERICAN
0.75 kg/1½ lb/1½ lb raspberries, strawberries or loganberries
1 kg/2 lb/4 cups castor sugar
½ bottle Certo
2 tablespoons/2 tablespoons/3 tablespoons lemon juice

Mash the fruit lightly with the sugar. Stirring occasionally, leave to stand for about 2½ hours in a warm kitchen, or until the sugar has dissolved. Add the Certo and lemon juice and stir for about 2 minutes. Pour into rigid polythene containers, leaving a 1-cm/½-inch headspace.
To freeze: Allow to stand for 48 hours in a warm kitchen, until set. Seal, label and freeze.
To thaw: Leave at room temperature for about 30 minutes when the jam should be ready for spreading.

Makes 1.5 kg/3 lb

Fruit purées

METRIC/IMPERIAL/AMERICAN
0.5 kg/1 lb/1 lb gooseberries or apricots
2–4 tablespoons/2–4 tablespoons/3–5 tablespoons water

Wash and top and tail the gooseberries or stone the apricots. Place in a saucepan with the water. Cover and cook very gently until pulpy, stirring occasionally to prevent the fruit sticking to the pan. Allow to cool, then sieve or liquidise.
Note: Soft fruits like raspberries and strawberries can be puréed by sieving without cooking and without the addition of any water.
To serve at once: Use as required in chosen recipe, adding sugar if necessary.
To freeze: Allow to cool completely, then pack in freezer boil-in-the-bags, foil bags, foil containers or rigid polythene containers. Seal, label and freeze.
To thaw: Thaw overnight in the refrigerator, for 6 hours at room temperature, or in hot water if frozen in a boil-in-the-bag.

Makes 300–450 ml/½–¾ pint/1¼–2 cups purée

Small choux buns

METRIC/IMPERIAL/AMERICAN
300 ml/½ pint/1¼ cups water
50 g/2 oz/¼ cup butter or margarine
150 g/5 oz/1¼ cups plain flour
2 large eggs

Place the water and butter or margarine in a saucepan and heat gently to melt the butter. Bring to the boil, remove from the heat and sieve the flour into the pan. Return to a moderate heat and beat well until the mixture forms a ball in the centre of the pan. Allow the mixture to cool a little. Beat the eggs well and, off the heat, gradually add them to the mixture in the pan, beating until smooth and glossy.

Place in a large nylon piping bag fitted with a star vegetable nozzle. Pipe small rosettes on baking sheets. Cook in the centre of a moderately hot oven (200°C, 400°F, Gas Mark 6) for 20 minutes until risen and crisp. Pierce the sides immediately to release the steam, and place in the switched-off oven for 20 minutes to dry off a little. Allow to cool on a cake rack.
To serve at once: The choux buns can now be filled and served as profiteroles (see page 45).
To freeze: Place the cooled buns on a baking sheet lined with greaseproof paper and open freeze. When solid, transfer to a polythene bag. Seal, label and return to the freezer.
To thaw: Unwrap and leave at room temperature for 1–2 hours. Serve as above.

Makes about 16

Smoked salmon rolls

METRIC/IMPERIAL/AMERICAN
1 small wholemeal loaf
100 g/4 oz/½ cup butter
100 g/4 oz/¼ lb smoked salmon
freshly ground black pepper
lemon juice
chopped parsley to garnish

Cut the crusts from the top, bottom, both ends and one side of the loaf. Lay the loaf on a bread board with the remaining crust on one side. Holding the crusted side in one hand, lightly spread the opposite side with butter. Slice off the buttered side as thinly as possible, cutting along the length of the loaf. Continue in this way until the loaf is finished. Discard the crust. Place a piece of smoked salmon on each slice of bread, leaving a 1-cm/½-inch border along one short side and narrow borders around the other sides. Sprinkle with black pepper and lemon juice. Roll up neatly from the 1-cm/½-inch border, spreading the final edge with a little extra butter if it does not stick.

To serve at once: Trim the rolls and cut into 2 or 3 equal pieces. Garnish with parsley.
To freeze: Pack the whole rolls interleaved with freezer layer tissue, in rigid containers. Seal, label and freeze.
To thaw: Leave to thaw for 30 minutes then trim as before. Cover with foil and leave for a further hour. Cut each roll into 2 or 3 pieces and serve garnished with parsley.

Makes approximately 8 whole rolls

Pâté pinwheels

METRIC/IMPERIAL/AMERICAN
1 large uncut white sandwich loaf
225 g/8 oz/1 cup butter or margarine, softened
225 g/8 oz/1 cup soft Continental liver sausage or soft pâté
GARNISH:
parsley sprig
stuffed green olives

Cut the crusts from the top, bottom, both ends and one side of the loaf. Holding the remaining crusted side in one hand, spread the opposite side thinly with the butter or margarine and pâté. Slice thinly along the length of the bread. Roll up tightly from the short end along the length of the slice. Continue in this way until all the bread is used up.

To serve at once: Trim the ends and cut each roll into 5 equal slices like a Swiss roll. Serve garnished with parsley and green olives.
To freeze: Place the whole rolls, interleaved with freezer layer tissue, in rigid polythene containers, or make into foil parcels, being careful not to squash them. Seal, label and freeze.
To thaw: Remove the required number from the freezer. After 1 hour trim the ends and cut into 5 equal slices as before. Arrange on a plate and cover with cling film. Leave to thaw completely.

Makes approximately 50 pinwheels

Pancakes

METRIC/IMPERIAL/AMERICAN
100 g/4 oz/1 cup plain flour
salt
1 egg
300 ml/½ pint/1¼ cups milk
oil for frying

Place the flour and salt in a bowl. Break in the egg and gradually add the milk to make a smooth batter.

Heat a little oil in a 20–23-cm/8–9-inch frying pan and pour in a little of the batter to make a very thin coating. Cook until the underside is golden brown. Either toss or turn over with a palette knife and cook this side for only a few seconds.
To serve at once: Serve with your favourite topping.
To freeze: Turn the pancakes on to kitchen paper on a cake rack to cool. Make into piles of four with greaseproof paper or freezer layer tissue between each pancake.

Wrap each pile in foil and make a parcel, or place in a plastic bag. Seal, label and freeze.
To thaw: Leave in parcels at room temperature for 2 hours.

Makes 7–8 pancakes

Asparagus rolls

METRIC/IMPERIAL/AMERICAN
1 (340-g/12-oz/12-oz) can green asparagus spears
175 g/6 oz/¾ cup butter
1 large wholemeal loaf

Drain the asparagus and turn on to a plate. Cut off the crust from the bottom, top, both ends and one side of the loaf. Lay the loaf on a bread board with the remaining crust on one side. Holding the crusted side in one hand, lightly spread the opposite side with butter. Slice off the buttered side as thinly as possible, cutting along the length of the loaf. Continue buttering and slicing until the loaf is finished. Discard the remaining crust. Cut each slice in half and place an asparagus spear on each half with a little of the tip showing. Roll up carefully. Trim off any extra stalk. If the roll does not hold together spread a little extra butter along the edge.
To serve at once: Keep covered with cling film until ready to use.
To freeze: Pack carefully, interleaved with freezer layer tissue, in rigid polythene containers. Seal, label and freeze.
To thaw: Arrange on a plate, cover with foil and leave for about 1 hour before serving.

Makes approximately 16 rolls

Soups and starters

Soups and pâtés are both very good candidates for the freezer and they are especially useful when planning a dinner party menu. They can be prepared well in advance and will require very little attention on the day. Frozen homemade stock can be used in the soups, making them even more delicious.

Tomato and cucumber soup

METRIC/IMPERIAL/AMERICAN
1 small onion, sliced
1 large cucumber, coarsely chopped
150 ml/¼ pint/⅔ cup chicken stock
1 (794-g/1-lb 12-oz/28-oz) can peeled tomatoes
salt
½ teaspoon paprika
pinch sugar
TO SERVE:
150 ml/¼ pint/⅔ cup natural yogurt
2 tablespoons/2 tablespoons/3 tablespoons single cream or creamy milk
chopped chives to garnish

Place the onion, cucumber and stock in a saucepan. Cover and simmer gently until the onion is soft. Sieve or liquidise with the tomatoes, salt, paprika and a pinch sugar. Chill well before serving.
To serve at once: Stir in the yogurt and single cream. Garnish each bowl with chopped chives.
To freeze: Do not add the yogurt or cream. Pour the tomato mixture into a rigid polythene container or foil bag. Allow to cool. Seal, label and freeze.
To thaw: Place in the refrigerator overnight as the soup should be served chilled. Finish as above.

Serves 6

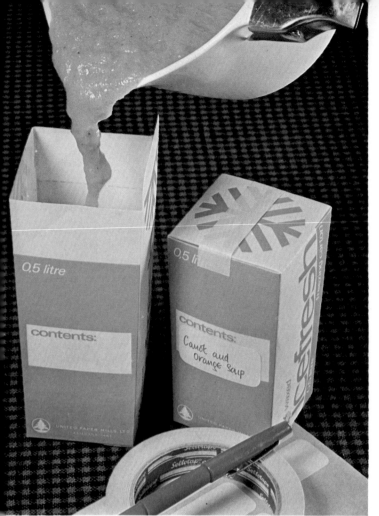

Carrot and orange soup

METRIC/IMPERIAL/AMERICAN
0.5 kg/1 lb/3 cups carrots, peeled and sliced
75 g/3 oz/¾ cup onion, finely chopped
750 ml/1¼ pints/3 cups chicken stock
salt and pepper
2 teaspoons sugar
bouquet garni
rind of ½ orange, thinly peeled
TO SERVE:
150 ml/¼ pint/⅔ cup single cream
croûtons and chopped parsley to garnish

Place the carrots, onion, stock, salt, pepper, sugar, bouquet garni and orange peel in a saucepan.
   Bring slowly to the boil, cover and simmer gently for 30 minutes, or until the vegetables are just tender. Remove the bouquet garni and orange rind and leave to cool a little before sieving or liquidising.
To serve at once: Add the cream to the soup and heat through gently, being careful not to allow it to boil. Adjust seasoning and serve sprinkled with fried bread croûtons and a little chopped parsley.
To freeze: Pour the sieved or liquidised soup into waxed cartons or rigid polythene containers without adding the cream. When quite cold, seal, label and freeze.
To thaw: Either thaw at room temperature for several hours or turn, frozen, into a saucepan and heat gently. Finish as above.

Serves 4–5

Cream of celery soup

METRIC/IMPERIAL/AMERICAN
1 large head celery
50 g/2 oz/½ cup onion, chopped
15 g/½ oz/1 tablespoon butter or margarine
600 ml/1 pint/2½ cups chicken stock
salt and pepper
TO SERVE:
300 ml/½ pint/1¼ cups milk
150 ml/¼ pint/⅔ cup single cream
25 g/1 oz/¼ cup Danish blue cheese, crumbled (optional)
celery leaves to garnish

Trim and wash the celery and chop roughly. Place in a saucepan with the onion and butter or margarine. Sauté very slowly for about 10 minutes without browning, then add the stock, salt and pepper. Cover and simmer gently for 30 minutes. Allow to cool before sieving or liquidising.
To serve at once: Return the soup to the saucepan, add the milk and heat through. Adjust seasoning. Just before serving, stir in the cream and the Danish blue cheese, if liked. Garnish with celery leaves.
To freeze: Pour the sieved or liquidised purée into a foil or waxed carton or a rigid polythene container. Allow to become quite cold then seal, label and freeze.
To thaw: Turn the purée into a pan and heat gently to thaw. When thawed, add the milk and finish as above.

Serves 4–6

Mushroom soup

METRIC/IMPERIAL/AMERICAN
25 g/1 oz/¼ cup onion, chopped
225 g/8 oz/2 cups mushrooms, washed and chopped
300 ml/½ pint/1¼ cups chicken stock
salt and pepper
TO SERVE:
15 g/½ oz/1 tablespoon butter
15 g/½ oz/2 tablespoons flour
300 ml/½ pint/1¼ cups milk
4 tablespoons/4 tablespoons/⅓ cup single cream
croûtons and chopped parsley to garnish

Place the onion and mushrooms in a saucepan with the stock and a little salt and pepper. Cover and simmer for 10 minutes. Allow to cool before sieving or liquidising.
To serve at once: Return the soup to the saucepan. Blend together the butter and flour to form a paste. Add to the soup in small pieces, then pour in the milk. Whisk until heated through. Adjust seasoning if necessary. Pour into soup bowls and add a little cream to each bowl. Sprinkle with croûtons and chopped parsley.
To freeze: Pour the sieved or liquidised mushroom mixture into a freezer container. Seal, label and freeze.
To thaw: Put the frozen soup into a saucepan with the milk and heat through gently. Add the butter and flour paste and whisk until thickened. Serve as above.

Serves 4

Cucumber vichyssoise

METRIC/IMPERIAL/AMERICAN
1 cucumber
50 g/2 oz/¼ cup butter
0.5 kg/1 lb/1 lb potatoes, peeled and diced
1 small onion, chopped
generous litre/2 pints/5 cups stock
TO SERVE:
300 ml/½ pint/1¼ cups single cream
salt and freshly ground pepper
150 ml/¼ pint/⅔ cup soured cream
chopped chives to garnish

Wash the cucumber but do not peel. Cut into cubes.
Melt the butter in a saucepan, add the vegetables, cover and cook gently for 10 minutes without browning. Add the stock, bring to the boil and simmer for 15–20 minutes or until the potatoes are tender. Sieve or liquidise.
To serve at once: Pour the soup into individual bowls and stir in the cream. Add salt and pepper to taste. Allow to chill thoroughly. Spoon a blob of soured cream into each bowl and serve sprinkled with chopped chives.
To freeze: Pour the sieved or liquidised soup into 1 or 2 freezer containers without the cream or seasoning. Seal, label and freeze.
To thaw: Defrost overnight in the refrigerator. Return to the liquidiser or heat gently to reblend. Stir in the cream and seasoning. Serve chilled with soured cream and chives as above.

Serves 6

Scampi with relish

METRIC/IMPERIAL/AMERICAN
0.5 kg/1 lb/2 cups tomato and onion relish, thawed (see page 12)
0.75 kg/1½ lb/1½ lb potatoes, peeled and sliced
salt and pepper
65 g/2½ oz/5 tablespoons butter
2 tablespoons/2 tablespoons/3 tablespoons milk
0.5 kg/1 lb/1 lb frozen scampi, thawed
150 ml/¼ pint/⅔ cup dry white wine
150 ml/¼ pint/⅔ cup water
1 small onion, sliced
bouquet garni
40 g/1½ oz/6 tablespoons plain flour
chopped parsley to garnish

Divide the relish between 6 individual flameproof dishes. Cook the potatoes in boiling salted water, drain and mash with 25 g/1 oz/2 tablespoons butter, the milk and salt and pepper.
　Simmer the scampi with the wine, water, onion and bouquet garni for 10–15 minutes. Drain well, reserving the stock. Divide the scampi between the dishes. Melt the remaining butter and stir in the flour. Gradually blend in the reserved stock. Bring to the boil, stirring, and cook for 1 minute. Season and pour over the scampi. Pipe potato around each dish. Place under a moderate grill for 10 minutes. Sprinkle with parsley.
Note: This recipe uses frozen ingredients and is not for refreezing.

Serves 4

Prawn and tomato cocktail

METRIC/IMPERIAL/AMERICAN
100 g/4 oz/½ cup tomato and onion relish, thawed (see page 12)
4 tablespoons/4 tablespoons/5 tablespoons mayonnaise
few drops Tabasco sauce
175 g/6 oz/1 cup frozen peeled prawns, thawed
1 small lettuce
paprika (optional)
4 fresh prawns or slices of lemon to garnish

Mix the tomato and onion relish with the mayonnaise and a few drops Tabasco sauce. Drain the thawed prawns on kitchen paper and then stir into the sauce.
　Shred the lettuce and use to line 4 ramekins or glasses suitable for a prawn cocktail. Divide the prawn mixture between the dishes and sprinkle with a little paprika if liked.
　Garnish each dish with a fresh prawn or a slice of lemon. Serve with toast.
Note: This recipe uses frozen ingredients and is not for refreezing.

Serves 4

Potted shrimps and mushrooms

METRIC/IMPERIAL/AMERICAN
50 g/2 oz/½ cup onion, finely chopped
350 g/12 oz/3 cups mushrooms, washed and chopped
350 g/12 oz/1½ cups unsalted butter
225 g/8 oz/1 cup peeled shrimps
50 g/2 oz/1 cup fresh white breadcrumbs
salt and pepper
juice of 1 lemon
1 tablespoon chopped parsley
shrimps and parsley sprigs to garnish

Place the onion and mushrooms in a pan with 100 g/4 oz/½ cup butter and heat gently to soften the onion. Stir in the shrimps, breadcrumbs, salt, pepper, lemon juice and chopped parsley. Divide the mixture between 4–6 freezerproof ramekins or foil dishes. Melt the remaining butter and allow to cool but do not leave until set. Spread over the dishes.
To serve at once: Chill before serving. Garnish each dish with a shrimp and a parsley sprig.
To freeze: Cover each dish with foil. Seal, label and freeze.
To thaw: Thaw for about 6 hours at room temperature or overnight in the refrigerator.

Serves 4–6

Smoked mackerel and cheese pâté

METRIC/IMPERIAL/AMERICAN
350 g/12 oz/¾ lb smoked mackerel
100 g/4 oz/½ cup cottage cheese, sieved
200 g/7 oz/scant cup butter
2 tablespoons/2 tablespoons/3 tablespoons creamed horseradish
juice of ½ lemon
3 tablespoons/3 tablespoons/4 tablespoons single cream
salt and freshly ground pepper
GARNISH:
slice of lemon
stuffed olives, sliced
parsley sprig

Remove the skin and bones from the fish and mash the flesh. Add the cottage cheese, 100 g/4 oz/½ cup melted butter and the remaining ingredients. Blend well. If liked, the pâté could be made smoother by blending in a liquidiser. Spoon the pâté into a serving dish, or a foil dish for freezing. Smooth the top. Melt the remaining butter and pour over the pâté. Arrange the lemon and olives in the centre and place in the refrigerator for the butter to set. Top with a sprig of parsley before serving.
To serve at once: Serve with French bread or melba toast.
To freeze: Cover, seal, label and freeze.
To thaw: Uncover and thaw at room temperature for 4–6 hours.

Serves 6

Cod's roe and tuna pâté

METRIC/IMPERIAL/AMERICAN
1 (198-g/7-oz/7-oz) can pressed cod roes
1 (198-g/7-oz/7-oz) can tuna fish
2 teaspoons grated onion
1 teaspoon anchovy essence
2 teaspoons lemon juice
freshly ground pepper
150 ml/¼ pint/⅔ cup soured cream
2 tablespoons/2 tablespoons/3 tablespoons chopped chives or parsley (optional)
salt
GARNISH:
lettuce leaves
twists of lemon

Place the cod roes and tuna in a bowl with the liquor from the cans. Mash well. Beat in the onion, anchovy essence, lemon juice, pepper and soured cream. There is no need to add salt at this stage as freezing will bring out the saltiness. Stir in the chives or parsley if liked.
To serve at once: Add a little salt if necessary. Arrange each portion on a bed of lettuce and garnish with a twist of lemon. Serve with melba toast.
To freeze: Pack in rigid polythene containers. Seal, label and freeze.
To thaw: Leave overnight in the refrigerator or at room temperature for about 4–6 hours. Serve as above.

Serves 8–12

Country pâté

METRIC/IMPERIAL/AMERICAN
150 ml/¼ pint/⅔ cup milk
15 g/½ oz/1 tablespoon butter
15 g/½ oz/2 tablespoons flour
100 g/4 oz/¼ lb lean belly of pork, derinded
100 g/4 oz/¼ lb pig's liver
1 small onion
1 small clove garlic
100 g/4 oz/½ cup sausagemeat
salt and pepper
1 bay leaf
3 black olives and 1 bay leaf to garnish

Heat the milk, butter and flour in a saucepan, whisking continuously until thickened. Reserve. Mince the pork twice with the liver, onion and garlic. Beat into the sauce with the sausagemeat and seasoning. Turn into a terrine, or a 0.5-kg/1-lb loaf tin for freezing. Place a bay leaf on top. Cover with foil and put in a roasting tin half filled with water. Cook in the centre of a moderate oven (180°C, 350°F, Gas Mark 4) for 1 hour. Allow to cool before turning out. Chill overnight.
To serve at once: Garnish with olives and a fresh bay leaf.
To freeze: If liked, cut into slices and separate with greaseproof paper. Wrap in foil. Seal, label and freeze.
To thaw: At room temperature for several hours.

Serves 6

Herbed chicken liver pâté

METRIC/IMPERIAL/AMERICAN
1 large onion, chopped
25 g/1 oz/2 tablespoons butter or margarine
1 tablespoon corn oil
0.5 kg/1 lb/1 lb chicken livers
¼ teaspoon chopped fresh tarragon
¼ teaspoon chopped fresh chives
pinch chopped fresh sage
1 teaspoon chopped fresh parsley
50 g/2 oz/¼ cup cream cheese
4 tablespoons/4 tablespoons/⅓ cup natural yogurt
garlic salt
freshly ground pepper
1 bay leaf to garnish

Place the onion, butter and oil in a saucepan and cook without browning until soft – about 10 minutes. Add the livers to the onion together with the chopped herbs. Mix well and cook for about 10 minutes.

Remove from the heat and sieve or blend in a liquidiser. Add the cream cheese and stir in the yogurt, garlic salt and pepper.

To serve at once: Transfer to a serving dish and chill to allow flavours to develop. Garnish with a bay leaf.
To freeze: Divide into 2 rigid polythene or foil containers, each one to serve 4 people. When quite cold, seal, label and freeze.
To thaw: Overnight in the refrigerator. Serve as before.

Serves 8

Liver and mushroom pâté

METRIC/IMPERIAL/AMERICAN
100 g/4 oz/1 cup onion, chopped
1 clove garlic, crushed
350 g/12 oz/¾ lb lambs' liver, sliced
100 g/4 oz/¼ lb streaky bacon, derinded and chopped
100 g/4 oz/1 cup mushrooms, washed and chopped
50 g/2 oz/¼ cup butter
40 g/1½ oz/6 tablespoons plain flour
300 ml/½ pint/1¼ cups milk
½ teaspoon dried basil
salt and pepper
cucumber, tomato and mustard and cress to garnish

Sauté the onion, garlic, liver, bacon and mushrooms in 15 g/½ oz/1 tablespoon butter for 10 minutes, then mince.

Melt the remaining butter, add the flour and gradually add the milk. Bring to the boil, stirring continuously, and cook for 2 minutes. Stir in the remaining ingredients and the minced mixture. Turn into a small terrine and cover. Place in a roasting tin half filled with water and cook in the centre of a moderate oven (160°C, 325°F, Gas Mark 3) for 1½ hours.
To serve at once: When cold, garnish and serve.
To freeze: When cold, remove the pâté from the dish, line the dish with foil, return the pâté and freeze. When frozen, lift the pâté out of the dish, wrap completely in foil and place in a polythene bag. Seal, label and return to the freezer.
To thaw: Overnight in the refrigerator.

Serves 6

Fish, meat and poultry

What could be better than to come home from a day at work or out with the family and have the dinner already prepared? Casseroles and stews are the usual favourites for freezing and this chapter provides some good ideas for these as well as giving you many other exciting recipes for freezing fish, meat and poultry.

Sole or plaice goujons

METRIC/IMPERIAL/AMERICAN
6–8 sole or plaice fillets
1 large egg
1 tablespoon water
salt and pepper
fresh breadcrumbs for coating
oil for deep frying

Wash the fillets and cut diagonally into strips about 1 cm/½ inch wide. Beat the egg with the water. Add salt and pepper. Spread the breadcrumbs on a sheet of greaseproof paper. Coat the fish strips in the egg and then toss in the breadcrumbs.
To serve at once: Place some of the goujons in a wire basket and fry for 1½ minutes in deep oil hot enough to brown a cube of bread in 30 seconds. Drain and keep hot while cooking a second batch. Serve with mixed salad and tartare sauce.
To freeze: Place the uncooked goujons on a baking sheet and open freeze. When frozen, place in a polythene bag. Seal, label and return to the freezer.
To thaw: Thaw for 30 minutes before frying. Heat deep fat until a cube of bread browns in 25 seconds. Place some of the goujons in the basket but do not overfill. Cook for about 1½ minutes, remove basket, allow fat to heat up again, then replace goujons for a further ½–1 minute to cook through and become crisp. Serve as above.

Serves 4

Individual fish pies

METRIC/IMPERIAL/AMERICAN
1 (396-g/14-oz/14-oz) packet frozen cod steaks, thawed
150 ml/¼ pint/⅔ cup milk
150 ml/¼ pint/⅔ cup water
40 g/1½ oz/3 tablespoons butter
1 small onion, grated
25 g/1 oz/¼ cup flour
2 teaspoons anchovy essence
1 tablespoon chopped fresh parsley
pepper and salt
1 (64-g/2¼-oz/2¼-oz) packet instant potato
prawns, slices of lemon and parsley sprigs to garnish

Cut the cod into cubes and poach in the milk and water for 10 minutes. Drain, reserving the liquor. Divide the fish between 4 scallop shells.

Sauté the onion in 25 g/1 oz/2 tablespoons butter for 2 minutes. Stir in the flour and reserved stock. Add the anchovy essence, parsley, pepper and salt. Pour over the fish.

Make up the instant potato, adding the remaining butter. Season to taste and pipe around each shell.

To serve at once: Place under a moderate grill for 10 minutes. Garnish and serve with a green salad.
To freeze: Open freeze then wrap in foil. Seal, label and return to the freezer.
To thaw: Unwrap and place the frozen pies in a hot oven (220°C, 425°F, Gas Mark 7) for 25 minutes.

Serves 4

Russian fish pie

METRIC/IMPERIAL/AMERICAN
0.5 kg/1 lb/1 lb cod
150 ml/¼ pint/⅔ cup milk
25 g/1 oz/2 tablespoons butter, melted
25 g/1 oz/¼ cup plain flour
salt and pepper
1 tablespoon chopped fresh parsley
1 (368-g/13-oz/13-oz) packet frozen puff pastry, thawed
½ egg, beaten
lettuce and watercress to garnish

Simmer the fish gently in the milk for 10–12 minutes. Drain and leave to cool, reserving the liquor.

Whisk together the butter, flour and reserved milk. Bring to the boil and continue whisking for 2–3 minutes. Flake the fish and fold into the sauce with salt, pepper and parsley.

Roll out the pastry to a 30-cm/12-inch square. Trim the edges. Place on a baking sheet and spoon the fish and sauce into the centre. Dampen the pastry edges and bring the corners together to make a square envelope. Seal and flute. Decorate with pastry leaves. Brush with beaten egg and cook in a hot oven (220°C, 425°F, Gas Mark 7) for 30–40 minutes.
To serve at once: Garnish with lettuce and watercress.
To freeze: Open freeze when cold, then wrap in foil. Seal, label and return to the freezer.
To thaw: Unwrap and place, frozen, on a baking sheet in a moderate oven (160°C, 325°F, Gas Mark 3) for 40–50 minutes.

Serves 4–6

Salmon fish cakes

METRIC/IMPERIAL/AMERICAN
350 g/12 oz/¾ lb old potatoes, peeled and boiled
50 g/2 oz/¼ cup butter
1 (212-g/7½-oz/7½-oz) can red salmon, drained and mashed
1 teaspoon anchovy essence
freshly ground white pepper
1 tablespoon chopped fresh parsley
1 egg, beaten
75 g/3 oz/¾ cup dried white breadcrumbs
1 tablespoon oil
parsley sprigs and twists of lemon to garnish

Mash the potatoes with half the butter. Mix with the salmon, anchovy essence, pepper and parsley. Divide into 6 equal portions and allow to cool a little.

On a lightly floured board, shape each portion into a round cake. Dip in the egg and then in the crumbs.

To serve at once: Fry in the remaining butter with the oil for 5 minutes on each side. Garnish with parsley sprigs and twists of lemon. Serve with parsley sauce.

To freeze: Place the uncooked fish cakes on a baking sheet and open freeze. When frozen, pack in a rigid polythene container, interleaved with freezer layer tissue. Seal, label and return to the freezer.

To thaw: Fry gently from frozen in melted butter and oil for about 15–20 minutes on each side.

Makes 6

Tuna pasties

METRIC/IMPERIAL/AMERICAN
90 g/3½ oz/scant ½ cup butter
300 g/11 oz/2¾ cups plain flour
150 ml/¼ pint/⅔ cup milk
1 (198-g/7-oz/7-oz) can tuna fish, drained and flaked
salt and pepper
2 teaspoons lemon juice
1 tablespoon chopped fresh chives
65 g/2½ oz/5 tablespooons lard
2–3 tablespoons/2–3 tablespoons/3–4 tablespoons water
1 small egg, beaten

Melt 25 g/1 oz/2 tablespoons butter and add 25 g/1 oz/¼ cup flour. Add the milk. Bring to the boil and cook for 2 minutes. Off the heat add the fish, salt, pepper, lemon and chives.

Rub the lard and remaining butter into the remaining flour and add water to make a dough. Knead lightly and roll out to a thickness of 3 mm/⅛ inch. Cut into 16 ovals – half slightly larger than the others. Divide the filling between the smaller ovals on a baking sheet. Cover with tops. Seal and crimp.

To serve at once: Brush with beaten egg and bake in a hot oven (220°C, 425°F, Gas Mark 7) for 25–30 minutes.

To freeze: Open freeze the unbrushed and uncooked pasties then wrap in foil. Seal, label and return to the freezer.

To thaw: Brush with egg and place, frozen, in a moderately hot oven (200°C, 400°F, Gas Mark 6) for 20 minutes. Reduce heat to 180°C, 350°F, Gas Mark 4 for 20 minutes.

Serves 4

Homemade beefburgers

METRIC/IMPERIAL/AMERICAN
0.5 kg/1 lb/2 cups minced beef
50 g/2 oz/1 cup brown breadcrumbs
1 small onion, grated
½ teaspoon made mustard
1 egg
1 teaspoon Worcestershire sauce
salt and pepper
fat for frying
TO FINISH:
4 tomatoes, sliced
1 onion, cut into rings
8 baps
8 cocktail sticks
1 lettuce

Mix together all the ingredients except the fat. Divide into eight 1-cm/½-inch thick cakes.
To serve at once: Cook the beefburgers in the fat over high heat for 1 minute on each side. Reduce the heat and cook for a further 6–8 minutes, turning once. Place a beefburger with slices of tomato and onion rings in each bap. Close with cocktail sticks and arrange on a bed of lettuce.
To freeze: Open freeze the beefburgers on a lined baking sheet, then pack in a polythene bag interleaved with freezer layer tissue. Seal, label and return to the freezer.
To thaw: Thaw for 1 hour, then fry and serve as above.

Makes 8

Pancake layer pie

METRIC/IMPERIAL/AMERICAN
25 g/1 oz/2 tablespoons butter or margarine
25 g/1 oz/¼ cup plain flour
300 ml/½ pint/1¼ cups milk
salt and pepper
5 pancakes, thawed (see page 13)
½ quantity meat sauce, thawed (see page 10)
TOPPING:
150 ml/¼ pint/⅔ cup natural yogurt
1 egg
25 g/1 oz/¼ cup Parmesan cheese, grated
GARNISH:
slice of tomato
mustard and cress

Melt the butter and stir in the flour. Gradually add the milk, bring to the boil, stirring, and cook for 2 minutes. Season.
In a round ovenproof dish (lined with foil if freezing) layer the pancakes, meat sauce and white sauce alternately, ending with a pancake. Beat the yogurt and egg together and pour over the surface. Sprinkle with the cheese.
To serve at once: Place the pie in the centre of a moderate oven (180°C, 350°F, Gas Mark 4) for about 35–40 minutes, until heated. Garnish and serve with French bread and beans.
To freeze: Cool then freeze. When solid remove the foil parcel from the dish, wrap, seal, label and return to freezer.
To thaw: Remove the foil, return the frozen pie to the original dish and place in a moderate oven (180°C, 350°F, Gas Mark 4) for 1¼ hours. Garnish and serve as above.

Serves 4

Meat and vegetable pasties

METRIC/IMPERIAL/AMERICAN
100 g/4 oz/1 cup potatoes, grated
50 g/2 oz/½ cup carrots, grated
25 g/1 oz/2 tablespoons butter
1 tablespoon plain flour
½ teaspoon dried mixed herbs
½ quantity meat sauce, thawed (see page 10)
SHORTCRUST PASTRY:
55 g/2 oz/¼ cup lard
55 g/2 oz/¼ cup butter
225 g/8 oz/2 cups plain flour, sifted
1 teaspoon salt
2 tablespoons/2 tablespoons/3 tablespoons cold water
beaten egg for brushing

Sauté the potatoes and carrots in the butter for 5 minutes. Stir in the flour, cook for 3 minutes. Add the herbs and meat sauce, cook for a further 5 minutes. Cool.

Rub the lard and butter into the flour and salt and add water to bind. Knead gently, then roll out as two 25-cm/10-inch circles. Divide the meat between the circles. Dampen the pastry edges and seal. Place on a baking sheet, brush with egg and bake in a hot oven (220°C, 425°F, Gas Mark 7) for 20–25 minutes. Serve with vegetables or cold with salad.

To freeze: Open freeze, wrap in foil and return to the freezer.
To thaw: To serve cold, thaw overnight in the refrigerator. To serve hot, unwrap and thaw overnight then place in a moderate oven (180°C, 350°F, Gas Mark 4) for 15–20 minutes.

Makes 2 pasties each serving 2

Sweet and sour meatballs

METRIC/IMPERIAL/AMERICAN
225 g/8 oz/1 cup minced beef
225 g/8 oz/1 cup beef sausagemeat
50 g/2 oz/½ cup onion, grated
½ teaspoon dried mixed herbs
2 teaspoons chopped parsley
salt and pepper
1 tablespoon seasoned flour
2 tablespoons/2 tablespoons/3 tablespoons oil
1 carrot, peeled and cut into small strips
450 ml/¾ pint/2 cups beef stock
4 tablespoons/4 tablespoons/⅓ cup malt vinegar
75 g/3 oz/6 tablespoons demerara sugar
1½ tablespoons/1½ tablespoons/2 tablespoons cornflour
1 teaspoon soy sauce

Mix together the mince, sausagemeat, onion, herbs and seasoning. Form into 18–20 balls and toss in the seasoned flour. Fry gently in the oil for 15–20 minutes, then drain well.

Place the carrot, stock, vinegar and sugar in a saucepan. Cover and cook slowly for 5 minutes. Blend the cornflour with the soy sauce and a little water. Stir in some of the hot liquid. Return to the pan and bring to the boil, stirring. Spoon the sauce over the meatballs and serve on noodles.

To freeze: Pack the cooled meatballs and sauce separately.
To thaw: Thaw the meatballs in the refrigerator overnight then place in a hot oven for 15 minutes. Heat the sauce in a pan.

Serves 4

Beef loaf

METRIC/IMPERIAL/AMERICAN
150 ml/¼ pint/⅔ cup beef stock
75 g/3 oz/1½ cups brown breadcrumbs
0.5 kg/1 lb/2 cups lean minced beef
50 g/2 oz/½ cup onion, finely chopped
¼ teaspoon dried thyme
50 g/2 oz/⅓ cup dried milk powder
1 egg, well beaten
2 teaspoons tomato purée
1 tablespoon chopped parsley
1 teaspoon Worcestershire sauce
tomato and mustard and cress to garnish

Mix together all the ingredients and place the mixture in the centre of a large piece of foil and form into a loaf shape. Seal the edges of the foil over the loaf.

Put the parcel in a tin half filled with water and cook in a moderately hot oven (200°C, 400°F, Gas Mark 6) for 1–1¼ hours.
To serve at once: Open the foil and drain off any liquid. If serving cold, wrap again in the foil so that the loaf keeps its shape while cooling. Garnish with mustard and cress and slices of tomato. Serve with potatoes sprinkled with parsley.
To freeze: Allow to cool completely, then wrap in fresh foil. Seal, label and freeze.
To thaw: Leave to thaw overnight in the refrigerator. To serve hot, unwrap and place in a moderate oven (180°C, 350°F, Gas Mark 4) for 20–30 minutes.

Serves 6–8

Individual shepherd's pies

METRIC/IMPERIAL/AMERICAN
1 quantity basic meat sauce (see page 10)
100 g/4 oz/1 cup carrot, grated
1 tablespoon flour
0.5 kg/1 lb/1 lb potatoes, peeled and boiled
salt and pepper
freshly ground nutmeg
parsley to garnish

Heat the meat sauce through to thaw, then stir in the grated carrot and flour. Cook for 10 minutes. Cool slightly, then divide between 6 ramekin dishes. For freezing, use freezerproof ramekins or line the dishes with foil.

Mash the potatoes with the salt, pepper and nutmeg. Pipe a whirl of potato on each ramekin.
To serve at once: Place in a moderately hot oven (190°C, 375°F, Gas Mark 5) for 25–30 minutes. Garnish with parsley and serve with grilled tomatoes.
To freeze: Open freeze before the final cooking. Pack freezerproof ramekins in polythene bags. If the ramekins have been lined with foil, lift out the solid pies and wrap in foil. Seal, label and return to the freezer.
To thaw: Unwrap freezerproof ramekins and thaw at room temperature for 30–50 minutes before placing in a moderately hot oven (200°C, 400°F, Gas Mark 6) for 30 minutes. Unwrap pies frozen in foil, replace in original ramekins and heat from frozen for 40 minutes in a moderately hot oven.

Serves 6

Basic beef stew

METRIC/IMPERIAL/AMERICAN
0.75 kg/1½ lb/1½ lb leg of beef
40 g/1½ oz/3 tablespoons dripping or cooking fat
175 g/6 oz/1½ cups onions, chopped
225 g/8 oz/1½ cups carrots, sliced
2 tablespoons/2 tablespoons/3 tablespoons potato flour
750 ml/1¼ pints/3 cups beef stock
salt and pepper
chopped parsley to garnish

Cut the meat into 2.5-cm/1-inch cubes, removing any fat and sinew. Heat the fat in a large saucepan or flameproof casserole and fry the meat fairly quickly to seal in the juices. When browned all over, remove the meat with a slotted spoon on to a plate.

Add the onions and carrots to the pan and fry over a moderate heat to lightly brown. The flavour and richness of the stew is determined at this stage so it should not be rushed.

Return the meat to the pan and stir in the potato flour, stock and seasoning. Cover and cook very slowly for 2½ hours.

To serve at once: Sprinkle with parsley and serve with new potatoes and peas.
To freeze: Allow to become quite cold, before spooning into 1 or 2 freezer containers. Seal, label and freeze.
To thaw: Allow to thaw at room temperature for about 2 hours, then turn into a saucepan and heat gently until ready for serving.

Serves 5–6

Pork and apricot curry

METRIC/IMPERIAL/AMERICAN
1 kg/2 lb/2 lb pork fillet, cubed
2 tablespoons/2 tablespoons/3 tablespoons oil
100 g/4 oz/1 cup onion, chopped
1 tablespoon curry powder
2 tablespoons/2 tablespoons/3 tablespoons flour
450 ml/¾ pint/2 cups stock
100 g/4 oz/⅔ cup dried or fresh apricots, stoned
1 small green pepper, deseeded and sliced
salt and pepper
chopped parsley to garnish

Brown the meat in the oil. Transfer to a plate. Add the onion to the pan and cook gently for 10 minutes. Stir in the curry powder and flour and cook for 2 minutes. Add the stock, bring to boil and simmer for 5 minutes. Remove from heat. Add apricots, pepper, meat and seasoning. Cover and simmer for 1 hour.

To serve at once: Garnish with parsley and serve with rice.
To freeze: When cold, remove the curry from the saucepan, wash the pan and line with foil. Return the food to the pan and freeze. Lift out the foil parcel, complete wrapping and place in a polythene bag. Seal, label and return to the freezer. The accompanying rice can be cooked in the usual way, cooled, and frozen in a heat-sealed freezer boil-in-the-bag.
To thaw: Unwrap the curry and return to the pan. Thaw for 3–4 hours then heat gently. Place bag of rice in a saucepan of cold water, bring to boil and simmer for 15–20 minutes.

Serves 6

Frikadeller

METRIC/IMPERIAL/AMERICAN
350 g/12 oz/1½ cups minced veal or beef
350 g/12 oz/1½ cups minced pork
1 tablespoon flour
½ teaspoon salt
¼ teaspoon pepper
¾ teaspoon ground allspice
1 small onion, grated
100 ml/4 fl oz/½ cup milk
1 egg, beaten
butter for frying
parsley sprigs to garnish

Mix the meats with the flour, salt, pepper, allspice and onion. Stir in the milk and beaten egg. Using 2 dessertspoons shape the mixture into oblongs.
To serve at once: Melt the butter in a frying pan and gently fry the frikadeller until brown on both sides – about 10–15 minutes altogether. Drain and garnish with parsley. Serve with red cabbage, or eat cold with coleslaw.
To freeze: Place the uncooked frikadeller well apart on a baking sheet and open freeze. When frozen, pack carefully in a rigid polythene container interleaving with waxed paper. Seal, label and return to the freezer.
To thaw: Remove the required number from the freezer, allow to thaw at room temperature for 2–3 hours and fry as before.

Serves 6

Stuffed shoulder of lamb

METRIC/IMPERIAL/AMERICAN
1 (1.5-kg/3-lb/3-lb) shoulder of lamb
watercress to garnish
STUFFING:
50 g/2 oz/½ cup onion, chopped
3 sticks celery, chopped
50 g/2 oz/¼ cup butter
100 g/4 oz/2 cups fresh breadcrumbs
50 g/2 oz/⅓ cup sultanas
salt and pepper
4 tablespoons/4 tablespoons/⅓ cup chopped parsley
¼ teaspoon grated nutmeg
1 egg, well beaten

Sauté the onion and celery in the butter until softened. Stir in the remaining stuffing ingredients.

Using a sharp, pointed knife, cut into the blade bone from the underside of the shoulder and ease the knife on each side to loosen the meat all round. Lift out the blade bone. Ease the knife round shoulder bone to loosen. Lift out the shoulder bone.
To serve at once: Fill the resulting cavities with stuffing and sew up with fine string. Cook in a moderate oven (190°C, 375°F, Gas Mark 5) for 1¼ hours. Untie and garnish.
To freeze: Freeze the stuffing in a polythene bag. Wrap the boned shoulder in foil and polythene. Seal, label and freeze.
To thaw: Thaw both lamb and stuffing, still wrapped, for 5–6 hours at room temperature. Stuff the lamb and cook as above.

Serves 6

Lamb and olive stew

METRIC/IMPERIAL/AMERICAN
0.75 kg/1½ lb/1½ lb lamb fillet, cubed
175 g/6 oz/3 cups onion, chopped
1 stock cube, crumbled
1 large clove garlic, crushed
1 (396-g/14-oz/14-oz) can peeled tomatoes
10 pimento-stuffed olives
2 tablespoons/2 tablespoons/3 tablespoons tomato purée
1 teaspoon sugar
salt and pepper

Place the meat in a saucepan with the onion and heat gently to extract the fat from the meat. Cook for 10 minutes, then add the remaining ingredients. Mix well and pour into a casserole which should be lined with foil for freezing.

Cover and cook in the centre of a moderate oven (180°C, 350°F, Gas Mark 4) for 1 hour if the casserole is to be frozen or 1½ hours if not. Serve sprinkled with a little parsley.

To freeze: Cool thoroughly and then freeze until solid. Remove the foil parcel from the casserole, wrap completely in foil and place in a polythene bag. Seal, label and return to the freezer. The parsley can be frozen separately in a polythene bag.

To thaw: Unwrap and return to the casserole. Either thaw at room temperature for several hours and then place in a moderate oven (160°C, 325°F, Gas Mark 3) for about 1 hour, or heat through from frozen in the oven for 2 hours. The frozen parsley can be crumbled over the casserole.

Serves 4

Lamb curry

METRIC/IMPERIAL/AMERICAN
1 quantity curry sauce (see page 11)
0.5 kg/1¼ lb/1¼ lb lamb fillet
1 small eating apple, peeled, cored and chopped
GARNISH:
parsley sprigs
lemon wedges

Thaw the curry sauce, if frozen. Trim the lamb and cut into 2.5-cm/1-inch cubes. Place in a saucepan and cook slowly in its own fat to seal in the meat juices. Transfer the meat to a plate. Drain the fat from the pan and clean the pan round with kitchen paper.

Place the curry sauce in the pan, and add the apple and meat. Mix well.

Cover and cook very slowly for 1¼–1½ hours until the meat is tender.

To serve at once: Garnish with parsley sprigs and lemon wedges. Serve with plain boiled rice, salted peanuts, poppadums, redcurrant jelly, sliced banana and chopped tomatoes.

To freeze: When cold, transfer the curry to a rigid polythene container. Seal, label and freeze.

To thaw: Leave in the refrigerator overnight and then heat through in a saucepan, or heat through gently from frozen. Serve as above.

Serves 4

Kidney special

METRIC/IMPERIAL/AMERICAN

8–10 lambs' kidneys
4 rashers streaky bacon, derinded
1 medium onion, sliced
225 g/8 oz/1½ cups carrots, sliced
25 g/1 oz/2 tablespoons butter
1 tablespoon plain flour
1 (396-g/14-oz/14-oz) can peeled tomatoes
1 teaspoon sugar
2 bay leaves
salt and pepper
¼ teaspoon dried thyme

Halve, skin and core the kidneys. Stretch each rasher of bacon with the back of a knife, cut in half and roll up neatly.

Sauté the bacon rolls, onion and carrots in the butter until golden brown. Keep on one side of the pan and add the kidneys to the other side. Cook gently for 3 minutes. Stir in the flour and the juice from the can of tomatoes, then add the tomatoes and the remaining ingredients. Simmer gently for about 15–20 minutes.

To serve at once: Serve with mashed potato.
To freeze: Cool quickly and transfer to suitable freezer container. Seal, label and freeze.
To thaw: Thaw completely either overnight in the refrigerator or at room temperature for 3–4 hours. Heat through gently and serve with mashed potatoes as before.

Serves 4

Rabbit pie

METRIC/IMPERIAL/AMERICAN

1 (0.5-kg/1¼-lb/1¼-lb) rabbit, prepared
1 small onion, sliced
750 ml/1¼ pints/3 cups water
1 tablespoon cornflour blended with 3 tablespoons/
 3 tablespoons/¼ cup milk
50 g/2 oz/¼ cup streaky bacon, cooked and chopped
SHORTCRUST PASTRY:
65 g/2½ oz/5 tablespoons margarine
65 g/2½ oz/5 tablespoons lard
275 g/10 oz/2½ cups plain flour
¼ teaspoon salt
parsley sprig to garnish

Place the rabbit, onion and water in a casserole with seasoning. Cover and cook in a moderate oven (180°C, 350°F, Gas Mark 4) for 1¼ hours. Strip the meat from the bones. Strain the liquor into a saucepan and add the cornflour and bacon. Bring to the boil, stirring. Cool. Make the pastry in the usual way (see page 28) and roll out just under half to line a 25-cm/10-inch foil plate. Arrange the rabbit and sauce in the centre. Roll out the remaining pastry and use as a lid. Cook in a hot oven (220°C, 425°F, Gas Mark 7) for 15 minutes, then at 190°C, 375°F, Gas Mark 5 for 25–30 minutes. Garnish and serve with salad.
To freeze: Open freeze when cold, then wrap in foil.
To thaw: Unwrap and place, frozen, in a moderate oven (180°C, 350°F, Gas Mark 4) for 40–45 minutes.

Serves 4–6

Chicken Kiev

METRIC/IMPERIAL/AMERICAN
100 g/4 oz/½ cup butter
3 tablespoons/3 tablespoons/¼ cup chopped parsley
2 cloves garlic, crushed
salt and pepper
4 chicken breasts
2 tablespoons/2 tablespoons/3 tablespoons seasoned flour
1 egg, beaten
fresh breadcrumbs for coating
oil for deep frying

Blend the butter with the parsley, garlic, salt and pepper and divide into 4 rolls. Wrap in foil and chill until hard.

Flatten the breasts slightly with a rolling pin, then put a roll of butter in the centre of each breast. Roll up the chicken meat, completely enclosing the butter. Sew up, leaving a length of thread at one end.

Toss in seasoned flour, then dip in beaten egg and then breadcrumbs. Repeat the egg-and-breadcrumbing. Chill before cooking.

To serve at once: Fry the chicken for about 10 minutes in deep oil hot enough to brown a cube of bread in 60 seconds.

To freeze: Place the uncooked crumbed breasts on a baking sheet, cover loosely with foil and open freeze. When frozen interleave with freezer film and pack in a polythene bag.

To thaw: Leave at room temperature for 6 hours. Cook as above.

Serves 4

Chicken parcels

METRIC/IMPERIAL/AMERICAN
1 medium onion, finely chopped
½ small green pepper, deseeded and sliced
25 g/1 oz/2 tablespoons butter
1 tablespoon plain flour
1 (227-g/8-oz/8-oz) can pineapple rings
1 tablespoon concentrated curry sauce
1 tablespoon chopped parsley
salt and pepper
4 chicken quarters

Sauté the onion and pepper in the butter for about 5 minutes. Stir in the flour and cook for 3 minutes.

Strain the juice from the pineapple and gradually stir it into the flour mixture, together with the curry sauce. Bring to the boil, stirring, and cook for 2 minutes. Halve the pineapple rings and add to the sauce with the parsley and seasoning.

Place each chicken quarter in the centre of a piece of foil and spoon the sauce equally over the chicken. Seal the foil to make parcels. Cook in the centre of a moderate oven (180°C, 350°F, Gas Mark 4) for 40–45 minutes.

To serve at once: Unwrap and serve with rice.

To freeze: Cool quickly and freeze, sealed and labelled, in the foil the chicken was cooked in. Overwrap in a polythene bag.

To thaw: Remove the polythene and place the frozen parcels in a moderate oven (180°C, 350°F, Gas Mark 4) for 1 hour. Serve as above.

Serves 4

Chicken cacciatore

METRIC/IMPERIAL/AMERICAN
4 chicken quarters
75 g/3 oz/6 tablespoons butter
1 large onion, chopped
1 medium green pepper, deseeded and chopped
1 clove garlic, crushed
1 (396-g/14-oz/14-oz) can peeled tomatoes
150 ml/¼ pint/⅔ cup chicken stock
salt and pepper
watercress to garnish

Sauté the chicken for 25–30 minutes in 50 g/2 oz/4 tablespoons butter. Meanwhile, gently sauté the onion, green pepper and garlic in the remaining butter for about 10 minutes. Stir in the tomatoes, stock and seasoning. Cover and simmer for 15 minutes.
To serve at once: Spoon the sauce over the chicken on a heated serving dish. Garnish with watercress and serve with rice sprinkled with chopped parsley.
To freeze: Spoon the sauce over the chicken in a foil dish. When cold, seal, label and freeze.
To thaw: In the refrigerator overnight, then place in a moderate oven (180°C, 350°F, Gas Mark 4) for 25–30 minutes.

Serves 4

Chicken lasagne

METRIC/IMPERIAL/AMERICAN
1 medium onion, chopped
75 g/3 oz/6 tablespoons butter
1 clove garlic, crushed
2 tablespoons/2 tablespoons/3 tablespoons plain flour
600 ml/1 pint/2½ cups chicken stock
225 g/8 oz/1⅓ cups cooked chicken, chopped
100 g/4 oz/1 cup mushrooms, washed and chopped
3 tablespoons/3 tablespoons/¼ cup tomato purée
1 teaspoon dried mixed herbs
8 uncooked sheets lasagne
25 g/1 oz/¼ cup flour
300 ml/½ pint/1¼ cups milk
25 g/1 oz/¼ cup Parmesan cheese, grated

Sauté the onion in 50 g/2 oz/¼ cup butter. Stir in the garlic, flour and stock. Bring to the boil, stirring. Off the heat add the chicken, mushrooms, tomato purée and herbs. Season. Arrange a layer of lasagne in a foil-lined ovenproof dish. Spread chicken sauce over. Repeat, finishing with sauce.
To serve at once: Melt remaining butter and stir in flour. Add milk. Cook for 2 minutes. Season, pour over lasagne. Sprinkle with cheese and place in a moderate oven for 45 minutes.
To freeze: Cool. Open freeze, then lift out of the dish, wrap in foil, seal, label and return to the freezer.
To thaw: Unwrap and return to the dish. Place in a moderate oven (180°C, 350°F, Gas Mark 4) for 1–1½ hours.

Serves 4–5

Vegetable dishes

Having a freezer will enable you to freeze down your homegrown produce and take advantage of good seasonal offers in the shops. Detailed instructions for blanching and freezing vegetables are given on page 8 so the recipes in this chapter concentrate on freezing cooked vegetable dishes and serving frozen vegetables in interesting ways.

Stuffed courgettes

METRIC/IMPERIAL/AMERICAN
8 large fresh courgettes
salt and pepper
100 g/4 oz/½ cup bacon, finely chopped
2 cloves garlic, crushed
1 large onion, chopped
2 tablespoons/2 tablespoons/3 tablespoons corn oil
225 g/8 oz/1 cup tomatoes, peeled and chopped
25 g/1 oz/¼ cup dry breadcrumbs
1 tablespoon chopped parsley to garnish

Wash and dry the courgettes, halve lengthways, scoop out the centre flesh, and chop. Cook the courgette boats in boiling salted water for 5 minutes. Drain. Sauté the bacon, garlic and onion in 1 tablespoon oil. Stir in tomatoes, courgette flesh and crumbs. Season and press this mixture into the courgette boats. Place in a foil dish, spoon the remaining oil over, cover with foil and cook in the centre of a moderately hot oven (190°C, 375°F, Gas Mark 5) for 15 minutes.
To serve at once: Serve hot or cold, garnished with parsley.
To freeze: Open freeze when cold, then transfer to a foil dish. Seal, label and return to the freezer.
To thaw: To serve cold, uncover and thaw at room temperature for about 3 hours. To serve hot, heat through from frozen in a moderately hot oven (200°C, 400°F, Gas Mark 6) for 15–20 minutes.

Serves 4–6

Savoury topped broccoli

METRIC/IMPERIAL/AMERICAN
0.5 kg/1 lb/1 lb broccoli, fresh or frozen
1 tablespoon oil
100 g/4 oz/1 cup onion, chopped
1 clove garlic, crushed (optional)
225 g/8 oz/½ lb tomatoes, peeled
1 tablespoon tomato purée
pinch sugar
salt and pepper
100 g/4 oz/½ cup ham, cut into strips
50 g/2 oz/½ cup Cheddar cheese, grated

Cook the broccoli in boiling salted water until just tender. This will take about 15 minutes for fresh broccoli or about half that time for frozen. Drain and keep hot in a serving dish.

Meanwhile, place the oil in a saucepan, add the onion and cook for about 10 minutes, until soft. Add the garlic, tomatoes, purée, sugar, salt and pepper and cook until fairly pulpy – about 10 minutes. Spread this mixture over the broccoli, arrange the ham over the top and sprinkle with cheese. Place under a moderate grill until the cheese has melted and the dish has heated through. Serve with French bread.

To freeze broccoli: See vegetable freezing chart on page 8.

Serves 3–4

Asparagus pancakes

METRIC/IMPERIAL/AMERICAN
300 ml/½ pint/1¼ cups pancake batter or 8 frozen pancakes (see page 13)
24 asparagus spears, fresh or frozen
40 g/1½ oz/3 tablespoons butter or margarine
25 g/1 oz/¼ cup plain flour
150 ml/¼ pint/⅔ cup milk
salt and pepper
25 g/1 oz/¼ cup Cheddar cheese, grated
parsley sprig to garnish

Make 8 thin pancakes as on page 13 or remove them from the freezer and thaw. Cook the asparagus in boiling salted water for 10–15 minutes. Drain well and place 3 spears in the centre of each pancake. Roll up and place the pancakes in a flameproof dish.

Melt the butter or margarine in a small saucepan and add the flour. Cook, stirring, for 1 minute, then remove from the heat and gradually blend in the milk. Return to the heat and bring to the boil, stirring continuously. Season to taste and pour the sauce over the pancakes. Sprinkle with the grated cheese. Place under a moderate grill to heat through slowly. Garnish with a parsley sprig.

To freeze asparagus: See vegetable freezing chart on page 8.

Serves 8 as a starter or 4 as a main course

Vegetable nut loaf

METRIC/IMPERIAL/AMERICAN
1 small onion, finely chopped
2 sticks celery, finely chopped
1 tablespoon oil
2 medium tomatoes, peeled, quartered and deseeded
1 small carrot, chopped
1 small parsnip, chopped
2 large slices brown bread
50 g/2 oz/generous ¼ cup salted peanuts
¼ teaspoon grated nutmeg
¼ teaspoon dried marjoram
salt and pepper
2 eggs, beaten
2 bay leaves
tomato and watercress to garnish

Cook the onion and celery in the oil, without browning, until soft. Mince together with the other vegetables, the bread and peanuts. Add the nutmeg, marjoram, salt, pepper and eggs. Mix well. Place in a well-greased 0.5-kg/1-lb deep foil dish. Smooth the surface and place 2 bay leaves on top. Cover with greased greaseproof paper and bake in the centre of a moderate oven (180°C, 350°F, Gas Mark 4) for 1 hour.
To serve at once: Garnish and serve cold.
To freeze: Allow to become cold, wrap in foil and place in a polythene bag. Seal, label and freeze.
To thaw: Thaw, wrapped, for about 3 hours.

Serves 4–6

Onion tart

METRIC/IMPERIAL/AMERICAN
75 g/3 oz/6 tablespoons butter
0.75 kg/1½ lb/1½ lb Spanish onions, thinly sliced
40 g/1½ oz/3 tablespoons lard
175 g/6 oz/1½ cups plain flour
1–2 tablespoons cold water
3 eggs
150 ml/¼ pint/⅔ cup single cream
5 tablespoons/5 tablespoons/6 tablespoons milk
salt and pepper
chopped parsley to garnish

Melt half the butter in a saucepan, add the onions and place dampened greaseproof paper on top. Cover with a lid and cook very gently until beginning to soften.
   Rub the remaining butter and the lard into the flour and add a little cold water to form a dough. Roll out the pastry to line an 18–20-cm/7–8-inch flan ring.
   Beat together the eggs, cream, milk and seasoning. Arrange the onions in the flan base and pour in the egg mixture. Cook in a hot oven (220°C, 425°F, Gas Mark 7) for 15–20 minutes, then at 190°C, 375°F, Gas Mark 5 for 20–30 minutes. Garnish.
To freeze: When cold, wrap in foil, seal, label and freeze.
To thaw: To serve cold, thaw at room temperature for about 4 hours. To serve hot, thaw and then place in the centre of a moderate oven (180°C, 350°F, Gas Mark 4) for about 15 minutes.

Serves 4

Desserts and gâteaux

Delicious homemade sorbets and ice creams can be made quickly and easily with the aid of your freezer, and are ideal ways of using up a glut of soft fruit. Attractive party desserts often require time and care so it is a great help to be able to make them at your leisure and then bring them out later as special-occasion surprises.

Satsuma sorbets

METRIC/IMPERIAL/AMERICAN
10 large satsumas
150 g/5 oz/⅔ cup castor sugar
1 lemon
1 small orange
1 egg white
mint leaves to decorate (optional)

Cut the tops off 8 satsumas to make small lids. Using a grapefruit knife, carefully scoop out all the flesh, being careful not to puncture the skins. Remove the pips from the flesh then liquidise the flesh and strain the juice into a saucepan. Add the sugar and heat gently to dissolve. Remove from the heat.

Grate the rind from the lemon and orange. Squeeze the juice from the lemon, orange and the 2 remaining satsumas. Add the rind and juice to the saucepan. Pour the mixture into a polythene container, leave until cold, then freeze until slushy – about 3–4 hours. Keep the 8 skins and lids in a polythene bag. Whisk the egg white until stiff then fold into the slushy juices. Place the satsuma shells in the hollows of 8 tartlet tins and fill with the mixture. Put the lids on top. Freeze until firm, then replace in polythene bags. Seal, label and return to the freezer.
To serve: Place in the refrigerator for 45 minutes–1 hour. Decorate with mint leaves if available.

Serves 8

Raspberry sorbet

METRIC/IMPERIAL/AMERICAN
225 g/8 oz/½ lb raspberries, fresh or frozen
100 g/4 oz/½ cup granulated sugar
300 ml/½ pint/1¼ cups water
2 tablespoons/2 tablespoons/3 tablespoons lemon juice
1 egg white
whole raspberries to decorate

Thaw the raspberries if frozen. Place the sugar and water in a saucepan and heat gently to dissolve the sugar. Bring to the boil then allow to become quite cold.

Sieve or liquidise and then strain the raspberries. Add the resulting purée to the sugar syrup. Stir in the lemon juice and mix well.

Pour into a polythene container, cover and freeze until slushy – about 3–4 hours. Whisk the egg white until stiff and fold into the mixture. Freeze until solid. If not required immediately, seal, label and return to the freezer.

To serve: Place in the refrigerator and thaw for just long enough to enable you to get out spoonfuls – 30 minutes–1 hour. Too much thawing will cause the sorbet to become slushy. Spoon into dishes and decorate with a few fresh or thawed frozen raspberries.

Serves 4   *Illustrated on the cover*

Strawberry ice cream

METRIC/IMPERIAL/AMERICAN
100 g/4 oz/½ cup granulated sugar
300 ml/½ pint/1¼ cups water
0.75 kg/1¾ lb/1¾ lb strawberries
1 egg white
150 ml/¼ pint/⅔ cup double cream
whole strawberries to decorate

Gently heat the sugar and water to dissolve the sugar, then bring to the boil. Boil steadily for 10 minutes then leave until cold. Sieve or liquidise the strawberries to give approximately 300 ml/½ pint/1¼ cups purée. Mix the syrup and strawberry purée together. Pour into a polythene container, cover and freeze until slushy – about 3–4 hours.

Whisk the egg white until stiff, mash down the fruit ice and fold in the egg white until evenly mixed. Freeze again until semi-solid – about 1 hour. Remove from the freezer.

Whip the cream until softly stiff. Mash down the fruit ice, mix in the cream and freeze until solid. If not required immediately, seal, label and return to the freezer.

To serve: Place in the refrigerator for 30 minutes to soften slightly. Spoon into dishes, decorate with a few whole strawberries and serve with fan wafers.

Note: Other fruit purées may be used in place of the strawberry purée.

Serves 6

Chocolate ice cream crunch

METRIC/IMPERIAL/AMERICAN
40 g/1½ oz/3 tablespoons butter
2 tablespoons/2 tablespoons/3 tablespoons castor sugar
100 g/4 oz/1 cup crushed gingernut biscuits
1 tablespoon cocoa powder
2 tablespoons/2 tablespoons/3 tablespoons water
300 ml/½ pint/1¼ cups double cream, whipped
1 tablespoon brandy (optional)
1 large egg white
100 ml/4 fl oz/½ cup double cream, whipped, to decorate

Line a 1-kg/2-lb loaf tin with greaseproof paper. Melt the butter and sugar in a saucepan. Remove from the heat and stir in the biscuit crumbs. Allow to cool. Press half the mixture into the base of the tin. Dissolve the cocoa powder in the water and add to the cream, straining if there are any lumps. Stir in the brandy, if used. Whisk the egg white until stiff and fold in. Pour the mixture on to the biscuits in the tin. Level the surface and spread the remaining crushed biscuits on top. Put into the freezer for about 2 hours to become firm. Turn out of the tin. If not required immediately, pack in a rigid polythene container, seal, label and return to the freezer.
To serve: Pipe whipped cream along each side to decorate. Allow to soften for 2–3 hours in the refrigerator if using after freezing for longer than the 2 hours.

Serves 6–8

Pineapple polls

METRIC/IMPERIAL/AMERICAN
1 frozen Arctic Roll
4 canned pineapple rings
2 egg whites
100 g/4 oz/½ cup castor sugar
angelica to decorate

Cut the Arctic Roll into 4 slices and place on an ovenproof plate. Drain the pineapple rings and place a ring on each slice of Arctic Roll.
 Whisk the egg whites until stiff. Gradually whisk in half the sugar and fold in the remainder. Divide the meringue evenly between the slices, carefully covering the pineapple and Arctic Roll and piling it up.
 Stick pointed pieces of angelica into the tops and sprinkle with a little extra castor sugar. Place towards the top of a hot oven (230°C, 450°F, Gas Mark 8) for a few minutes to brown lightly. Serve at once.
Note: This recipe is for using frozen ingredients, not for freezing when cooked.

Serves 4

Lemon mousse

METRIC/IMPERIAL/AMERICAN
3 eggs, separated
175 g/6 oz/¾ cup castor sugar
3 large lemons
15 g/½ oz/2 envelopes gelatine
3 tablespoons/3 tablespoons/¼ cup water
DECORATION:
100 ml/4 fl oz/½ cup double cream, whipped
twist of lemon
angelica

Place the egg yolks with the sugar in a bowl over a pan of hot water. Whisk until light and fluffy and the whisk leaves a trail. Remove the bowl and continue whisking until the mixture has cooled. Stir in the grated rind from 2 lemons and the juice from three. Dissolve the gelatine in the water and stir into the lemon mixture. Put in a cool place. When the mixture is on the point of setting, whisk the egg whites until stiff and fold in evenly.
To serve at once: Pour into a fluted mould and chill well. When set, turn out and decorate with piped cream, a twist of lemon and small pieces of angelica.
To freeze: Pour into a polythene fluted mould. Freeze until solid then cover with a lid or foil. Seal, label and return to the freezer.
To thaw: Dip the mould into hot water for 30 seconds then turn out and leave uncovered at room temperature for 1–2 hours. Decorate as above.

Serves 4–6

Apricot soufflé

METRIC/IMPERIAL/AMERICAN
100 g/4 oz/¾ cup dried apricots, soaked overnight
8 tablespoons/8 tablespoons/⅔ cup water
1 orange jelly
4 large eggs, separated
100 g/4 oz/½ cup castor sugar
5 tablespoons/5 tablespoons/6 tablespoons apricot brandy
150 ml/¼ pint/⅔ cup double cream, lightly whipped
DECORATION:
100 ml/4 fl oz/½ cup double cream, whipped
1–2 tablespoons crushed sweet biscuits

Tie a band of foil around the outside of a 15-cm/6-inch freezerproof soufflé dish to extend 7.5 cm/3 inches above the rim. Cook the apricots gently in three-quarters of the water until pulpy. Sieve or liquidise. Dissolve the jelly very gently in the remaining water. Cool.
  Whisk the egg yolks and sugar in a large bowl over boiling water until the whisk leaves a trail. Remove from the heat and whisk until cool. Stir in the apricots, jelly and apricot brandy, then add the cream. Whisk the egg whites until stiff and fold in. Pour into the prepared dish. Chill until firm, then remove the collar. Decorate as illustrated.
To serve at once: The soufflé is now ready for serving.
To freeze: Open freeze, then pack into a rigid polythene container, seal, label and return to the freezer.
To thaw: Place in the refrigerator overnight.

Serves 4–6

Pears in ginger syrup

METRIC/IMPERIAL/AMERICAN
0.5 kg/1 lb/1 lb cooking pears
juice of 1 lemon
300 ml/½ pint/1¼ cups water
75 g/3 oz/6 tablespoons demerara sugar
3 tablespoons/3 tablespoons/¼ cup chopped stem ginger

Peel the pears, core and cut into quarters. Sprinkle with lemon juice to prevent discoloration. Place the water and sugar in a saucepan. Heat gently to dissolve the sugar, add the pears and cook gently until just tender – about 10 minutes. Test with a skewer.

Using a slotted spoon, transfer the pears to a serving dish or a container suitable for the freezer. Reduce the syrup a little over high heat and stir in the chopped ginger. Allow the syrup to cool before pouring over the pears.

To serve at once: Chill well and serve with ice cream or cream.
To freeze: Allow to cool, making sure the pears are completely covered by the syrup. Seal, label and freeze.
Thaw: At room temperature for 2–3 hours.

Serves 3–4

Summer pudding

METRIC/IMPERIAL/AMERICAN
50 g/2 oz/¼ cup castor sugar
4 tablespoons/4 tablespoons/⅓ cup water
0.5–0.75 kg/1–1½ lb/1–1½ lb raspberries and blackcurrants
1 small stale loaf, sliced
150 ml/¼ pint/⅔ cup double cream, whipped, to decorate

Gently heat the sugar and water to dissolve the sugar. Add the fruit, reserving a little of each for decoration. Simmer gently for 5 minutes.

Remove the crusts from the bread and use as many slices as necessary to line the base and sides of a 0.5-kg/1-lb foil basin with a sealing lid. Add half the fruit, place a slice of bread in the centre, add the remaining fruit and cover with more bread. Ease on the lid, place a 1-kg/2-lb weight on top and put in the refrigerator overnight.

To serve at once: Turn the pudding out, place the reserved fruit on top and pipe whipped cream around the edge.
To freeze: Leave in the foil basin. Pack in a polythene bag. Seal, label and freeze. Freeze the reserved fruit separately.
To thaw: Leave in the basin and thaw overnight in the refrigerator or for 6 hours at room temperature. Turn out and decorate as above, using the thawed reserved fruit.

Serves 4–5

Hazelnut Pavlova

METRIC/IMPERIAL/AMERICAN
3 large egg whites
175 g/6 oz/1⅓ cups icing sugar, sifted
50 g/2 oz/⅓ cup hazelnuts, roasted, skinned and chopped
1 teaspoon malt vinegar
¼ teaspoon vanilla essence
300 ml/½ pint/1¼ cups double cream, lightly whipped
300 ml/½ pint/1¼ cups apricot purée (see page 15)
2 tablespoons/2 tablespoons/3 tablespoons water
40 g/1½ oz/3 tablespoons castor sugar

Whisk the egg whites until stiff, and gradually whisk in half the icing sugar. Fold in the remainder with the hazelnuts, vinegar and vanilla. Spread in a 23-cm/9-inch circle on a sheet of non-stick paper on a baking sheet. Make a slight hollow in the centre. Bake for 1 hour in a very cool oven (120°C, 250°F, Gas Mark ½). Switch off oven and leave meringue for 1 hour.
To serve at once: Pile the cream into the centre. Decorate the edge with piped rosettes of cream, if liked. Gently heat the apricot purée with the water and sugar. Serve hot or cold.
To freeze: Wrap the cooled pavlova in foil, being careful not to squash it. Seal, label and freeze. Freeze the sauce separately.
To thaw: Unwrap the pavlova and leave at room temperature for 1 hour. Finish as before. Thaw the sauce at room temperature for 1–2 hours or heat gently from frozen. Place the rosettes, still frozen, on the pavlova.

Serves 6–8

Apricot linzertorte

METRIC/IMPERIAL/AMERICAN
75 g/3 oz/6 tablespoons granulated sugar
0.5 kg/1 lb/1 lb fresh apricots, halved and stoned
100 g/4 oz/½ cup butter
75 g/3 oz/6 tablespoons castor sugar
1 egg
175 g/6 oz/1½ cups plain flour
50 g/2 oz/½ cup ground hazelnuts
1 teaspoon ground cinnamon
pinch salt
1 tablespoon cornflour

Dissolve sugar in 2 tablespoons/2 tablespoons/3 tablespoons water. Add apricots and simmer for 15–20 minutes. Drain, reserve syrup. Add the butter, sugar and egg to the flour, nuts, cinnamon and salt, and knead. Chill.

Make the apricot syrup up to 150 ml/¼ pint/⅔ cup with water. Blend the cornflour with a little of this, place all in a pan. Heat, stirring, until thickened. Add apricots and cool.

Roll out two-thirds of the pastry to line a 20-cm/8-inch flan tin or foil tart case. Fill with the apricots and glaze. Roll out the remaining pastry, cut into strips and arrange over the top.
To serve at once: Bake at 220°C, 425°F, Gas Mark 7 for 10 minutes, then 180°C, 350°F, Gas Mark 4 for 25 minutes.
To freeze: Uncooked in a polythene bag. Seal, label and freeze.
To thaw: Unwrap and bake from frozen in a hot oven as above, cook for 20 minutes before reducing heat for further 30 minutes.

Serves 6

Crêpes Suzette

METRIC/IMPERIAL/AMERICAN
300 ml/½ pint/1¼ cups pancake batter or 8 thin frozen pancakes, thawed (see page 13)
20 sugar lumps
1 large orange
75 g/3 oz/6 tablespoons unsalted butter
few glacé cherries
2 tablespoons/2 tablespoons/3 tablespoons Cointreau or Grand Marnier
2 tablespoons/2 tablespoons/3 tablespoons demerara sugar
1 tablespoon brandy

Make up the batter and cook as in the basic recipe on page 13, or remove 8 pancakes from the freezer and allow to thaw.

Rub the sugar lumps on the orange rind to remove the zest and oil from the skin. Squeeze the juice from the orange. Place the butter, sugar lumps and orange juice in a frying pan or flambé dish. Heat gently to melt the butter and dissolve the sugar.

Fold the pancakes in four and add to the pan with the cherries. Stir in the liqueur and sprinkle with the sugar. Heat through for a few minutes.

Heat a tablespoon, pour in the brandy and ignite. Pour over the pancakes and serve at once.
Note: This recipe uses frozen ingredients and is not for refreezing.

Serves 4

Profiteroles with chocolate sauce

METRIC/IMPERIAL/AMERICAN
1 quantity choux pastry or 12 choux buns, thawed (see page 15)
100 g/4 oz/⅔ cup plain chocolate
25 g/1 oz/2 tablespoons castor sugar
150 ml/¼ pint/⅔ cup hot water
pinch salt
50 g/2 oz/¼ cup butter, diced
300 ml/½ pint/1¼ cups double cream, whipped

Make up the pastry and cook as page 15, or remove 12 buns from the freezer and allow to thaw.

Heat the chocolate gently with the sugar and hot water. When melted, stir in the salt. Simmer gently until reduced a little. Add the butter.
To serve at once: Pipe the cream into the buns. Stack the buns in a serving dish and pour some of the sauce over them. Serve the remaining sauce separately.
To freeze: If using frozen choux buns, do not refreeze. If making up the buns, freeze them unfilled, as on page 15. Freeze chocolate sauce in a freezer boil-in-the-bag.
To thaw: See page 15 for thawing the buns. Dip the bag of frozen sauce into boiling water until melted. Fill and finish the buns as above.

Serves 4

Special mince pies

METRIC/IMPERIAL/AMERICAN
175 g/6 oz/¾ cup butter, softened
100 g/4 oz/½ cup castor sugar
2 egg yolks
275 g/10 oz/2½ cups plain flour, sifted
0.5 kg/1 lb/2 cups mincemeat
grated rind of 1 lemon
1 small eating apple, peeled and grated
2 tablespoons/2 tablespoons/3 tablespoons brandy
icing sugar for dredging

Add the butter, sugar and egg yolks to the flour. Knead lightly to form a dough. Cover and chill. Mix together the mincemeat, lemon rind, apple and brandy. Roll out the pastry to a thickness of 5 mm/¼ inch and cut out 18 fluted rounds to line 18 greased tartlet tins. Divide the filling between the cases. Cut 18 slightly smaller rounds from the remaining pastry and use as tops. Flute the edges and cut 2 small holes in each top. Cook in a moderately hot oven (200°C, 400°F, Gas Mark 6) for 15 minutes.
To serve at once: Sprinkle with icing sugar and serve with lightly whipped cream flavoured with cinnamon.
To freeze: When cold, freeze in a rigid polythene container.
To thaw: Place, still frozen, in a moderate oven (180°C, 350°F, Gas Mark 4) for 25 minutes. Sprinkle with icing sugar and serve with cream as above.

Makes 18

Orange cream gâteau

METRIC/IMPERIAL/AMERICAN
3 eggs
75 g/3 oz/6 tablespoons castor sugar
75 g/3 oz/¾ cup plain flour, sifted
¼ teaspoon salt
3 oranges
300 ml/½ pint/1¼ cups double cream, lightly whipped

Line the base of a 20-cm/8-inch round cake tin with greaseproof paper and oil lightly.
  Whisk together the eggs and sugar in a large bowl over a pan of boiling water, until the whisk leaves a trail. Remove the bowl and continue to whisk until cool. Using a metal spoon, quickly fold in the flour and salt. Pour into the tin and cook in the centre of a moderate oven (180°C, 350°F, Gas Mark 4) for 25–30 minutes. Cool on a wire rack.
To serve at once: Cut in half horizontally. Reserve 1 slice of orange for decoration. Peel 2 oranges, chop the flesh and place on the base sponge. If liked, frozen orange slices can be thawed and cut up for this. Place the sponge on a serving dish and put the top half in position. Pour over the juice from the third orange, spread cream all over the cake and pipe rosettes of cream around the edge. Twist reserved slice and place on.
To freeze: Wrap the cooled cooked sponge without filling or decoration in a polythene bag or in foil. Seal, label and freeze.
To thaw: Thaw, still wrapped, for 1 hour at room temperature. Unwrap and finish as above.

Makes 1 (20-cm/8-inch) gâteau

Raspberry gâteau

METRIC/IMPERIAL/AMERICAN
3 eggs
75 g/3 oz/6 tablespoons castor sugar
75 g/3 oz/¾ cup plain flour, sifted
¼ teaspoon salt
2 tablespoons/2 tablespoons/3 tablespoons raspberry jam
300 ml/½ pint/1¼ cups double cream, whipped
0.5 kg/1 lb/1 lb fresh or frozen raspberries
2 tablespoons/2 tablespoons/3 tablespoons sherry
3 tablespoons/3 tablespoons/¼ cup redcurrant jelly

Line the base of a 20-cm/8-inch round cake tin with greaseproof paper and lightly oil.
Whisk together the eggs and sugar in a bowl over a pan of boiling water until the whisk leaves a trail. Remove the bowl and whisk until cool.
Using a metal spoon quickly fold in the flour and salt. Pour into the tin and cook in the centre of a moderate oven (180°C, 350°F, Gas Mark 4) for 25–30 minutes. Cool on a wire rack.
To serve at once: Cut in half and spread the base with the raspberry jam, a third of the cream and then half the raspberries. Cover with the top sponge and sprinkle with sherry. Decorate with the remaining cream and raspberries. Brush warmed redcurrant jelly over the raspberries.
To freeze: Place the undecorated sponge in a polythene bag or wrap in foil. Seal, label and freeze.
To thaw: Leave, wrapped, at room temperature for 1 hour.

Makes 1 (20-cm/8-inch) gâteau

Orange and lemon pudding

METRIC/IMPERIAL/AMERICAN
100 g/4 oz/½ cup granulated sugar
50 g/2 oz/½ cup plain flour
25 g/1 oz/2 tablespoons butter or margarine
1 orange
1 lemon
3 eggs, separated
450 ml/¾ pint/2 cups milk

Mix together the sugar, flour and butter or margarine in a bowl. Add the grated rinds and juice from the orange and lemon. Beat well. Add the egg yolks to the bowl, stir in the milk and mix well together. Whisk the egg whites until stiff and fold into the mixture.
Pour into a 1-litre/1½-pint/2-pint foil dish. Place in a roasting tin half filled with water and bake in the centre of a moderate oven (180°C, 350°F, Gas Mark 4) for 45 minutes until risen and brown.
To serve at once: Serve hot.
To freeze: Allow to cool then cover, seal, label and freeze.
To thaw: Uncover and place, still frozen, in a cool oven (150°C, 300°F, Gas Mark 2) for 20 minutes.

Serves 6

Apple bread and butter pudding

METRIC/IMPERIAL/AMERICAN
6 large slices bread, crusts removed
40 g/1½ oz/3 tablespoons butter, softened
75 g/3 oz/½ cup raisins
1 large cooking apple, peeled and sliced, or 175 g/6 oz/1½ cups frozen apple slices, thawed
40 g/1½ oz/3 tablespoons castor sugar
2 large eggs
450 ml/¾ pint/2 cups milk

Spread the bread with the butter and cut each slice into three. Place a layer of the bread pieces, buttered side up, in a 1.25-litre/2-pint/2½-pint ovenproof dish which must be freezerproof if it is to be frozen. Sprinkle with raisins, apple and sugar. Continue layering until the ingredients are used up.

Beat the eggs and milk together and strain over the bread. Allow to stand for 20 minutes then cook towards the bottom of a moderately hot oven (190°C, 375°F, Gas Mark 5) for 30 minutes.
To serve at once: Serve hot with cream.
To freeze: Cool, wrap in foil, seal, label and freeze.
To thaw: Leave, wrapped, at room temperature for 1 hour then place uncovered in a moderate oven (180°C, 350°F, Gas Mark 4) for 30–40 minutes.

Serves 4–6

Pineapple pudding

METRIC/IMPERIAL/AMERICAN
2 tablespoons/2 tablespoons/3 tablespoons cornflour
1 (376-g/13¼-oz/13¼-oz) can crushed pineapple
3 tablespoons/3 tablespoons/¼ cup marmalade
110 g/4 oz/½ cup soft margarine
110 g/4 oz/½ cup castor sugar
2 large eggs
110 g/4 oz/1 cup self-raising flour
grated rind of 1 orange

Mix the cornflour with a little juice from the pineapple. Heat the pineapple gently with the marmalade. Add the cornflour and stir until thickened. Spoon into a 1.25-litre/2-pint/2½-pint ovenproof dish. Use a foil or other freezerproof dish if the pudding is to be frozen. Cool.

Beat together the margarine, sugar, eggs, flour and orange rind for 2 minutes. Spread over the pineapple.
To serve at once: Cook in the centre of a moderately hot oven (190°C, 375°F, Gas Mark 5) for 35–40 minutes; cover up with foil, if necessary, to prevent overbrowning. Turn out and serve hot or cold.
To freeze: Place the uncooked pudding in the freezer. When frozen, cover with a lid or foil. Seal, label and return to the freezer.
To thaw: Remove lid but cover loosely with foil. Place, still frozen, in a moderate oven (180°C, 350°F, Gas Mark 4) for 1¾ hours. Remove foil for the last 15 minutes.

Serves 6–8

Upside-down pudding

METRIC/IMPERIAL/AMERICAN
1 (410-g/15-oz/15-oz) can pear halves or 0.5 kg/1 lb/1 lb frozen pear halves in syrup, thawed
2 tablespoons/2 tablespoons/3 tablespoons cocoa powder
40 g/1½ oz/3 tablespoons granulated sugar
110 g/4 oz/½ cup soft margarine
110 g/4 oz/½ cup castor sugar
175 g/6 oz/1½ cups self-raising flour
1½ teaspoons baking powder
2 tablespoons/2 tablespoons/3 tablespoons cocoa powder
2 large eggs
1 teaspoon vanilla essence
100 ml/4 fl oz/½ cup double cream, whipped, to decorate

Strain the juice or sugar syrup from the pears into a saucepan. Blend a little juice with the cocoa and strain back into the pan. Add the sugar, heat gently to dissolve, then boil for 15 minutes to reduce by half.

Arrange the pears in the base of a 23-cm/9-inch round foil flan dish. Pour the chocolate sauce over and allow to cool.

Beat the remaining ingredients together for 2 minutes and spread over the pears. Bake in the centre of a moderate oven (180°C, 350°F, Gas Mark 4) for 40–45 minutes. Turn out and cool on a wire rack.

To serve at once: Decorate with piped rosettes of cream.
To freeze: Cover with foil, seal, label and freeze.
To thaw: Overnight in the refrigerator. Decorate as above.

Serves 6–8

Caramel oranges

METRIC/IMPERIAL/AMERICAN
150 g/5 oz/⅔ cup castor sugar
300 ml/½ pint/1¼ cups water
5 medium oranges

Dissolve the sugar in half the water by heating gently, then bring to the boil and cook steadily until the mixture begins to caramelise and turn brown. Do not allow the mixture to become too dark. Plunge the base of the pan into warm water to stop the cooking. Leave to cool a little, then pour in the remaining water. Heat gently to dissolve the caramel. Remove the pan from the heat.

Using a sharp knife, carefully cut off the skin and pith from the oranges. Cut in between the membranes to remove the orange segments.

To serve at once: Arrange the orange segments in individual dishes or on a large serving dish and pour the cooled caramel over. Chill before serving. Serve with tuilles (see page 61).
To freeze: Place the orange segments in a rigid polythene container and pour the caramel over the fruit. Seal, label and freeze.
To thaw: Thaw overnight in the refrigerator or for about 2 hours at room temperature.

Serves 5

Party entertaining

No longer need you hesitate before giving a party for fear of all the last-minute food preparation. With your freezer to help, you can spread the work over several weeks and, in fact, be far more ambitious than before. Pizzas, galantines, mousses and risottos can take the place of cheese and French bread to the delight of your guests, and without you being too exhausted to enjoy the festivities.

Pizza

METRIC/IMPERIAL/AMERICAN
1 quantity basic white bread dough (see page 62)
2 large onions, chopped
1 tablespoon corn oil
2 (396-g/14-oz/14-oz) cans peeled tomatoes
225 g/8 oz/2 cups mushrooms, chopped
1 teaspoon salt
1 tablespoon sugar
1 (141-g/5-oz/5-oz) can tomato purée
1 tablespoon dried oregano
75 g/3 oz/¾ cup Cheddar cheese, grated
25 g/1 oz/¼ cup Parmesan cheese, grated
anchovies, olives and chopped parsley to garnish

Line 2 greased 23 × 28-cm/9 × 11-inch baking sheets with the dough.
   Sauté the onions in the oil in a large saucepan for 10 minutes. Add the tomatoes, mushrooms, salt, sugar, tomato purée and oregano. Mix well and cook uncovered for 20 minutes until thick. Cool and spread over the dough bases. Sprinkle with the cheeses. Leave to rise for about 10 minutes then place in a hot oven (230°C, 450°F, Gas Mark 8) for 15 minutes. Arrange anchovies and olives over the top. Sprinkle with parsley.
To freeze: Cut the cooled, ungarnished pizza into portions. Open freeze, then wrap in foil and a polythene bag. Seal, label and return to the freezer.
To thaw: Unwrap and place in a hot oven (220°C, 425°F, Gas Mark 7) for 10 minutes. Garnish as above.

Makes 2 pizzas, each serving 6

Savoury ham quiche

METRIC/IMPERIAL/AMERICAN
110 g/4 oz/½ cup lard
225 g/8 oz/2 cups plain flour
½ teaspoon salt
2 tablespoons/2 tablespoons/3 tablespoons cold water
100 g/4 oz/1 cup onion, minced or grated
350 g/12 oz/1½ cups ham or cooked bacon, minced
4 eggs
2 teaspoons dried mixed herbs
1 tablespoon chopped parsley
pepper
300 ml/½ pint/1¼ cups milk

Rub the lard into the flour, add the salt and enough cold water to make a dough. Roll out to line a 28 × 18-cm/11 × 7-inch rectangular tin. Mix the onion with the ham or bacon and spread over the pastry case. Beat the eggs, herbs, pepper and milk together. Pour over the ham and bake in the centre of a moderately hot oven (190°C, 375°F, Gas Mark 5) for 1 hour.
To serve at once: Serve hot or cold with a salad.
To freeze: Open freeze, then wrap in foil, seal, label and return to the freezer.
To thaw: To serve cold, unwrap and thaw at room temperature for 2 hours. To serve hot, unwrap and place, still frozen, in a moderate oven (180°C, 350°F, Gas Mark 4) for 1 hour.

Serves 6 as a main meal, 12 as part of a buffet, 30 for a cocktail party

Galantine of chicken

METRIC/IMPERIAL/AMERICAN
50 g/2 oz/½ cup onion, chopped
50 g/2 oz/½ cup celery, chopped
25 g/1 oz/2 tablespoons butter
50 g/2 oz/1 cup fresh white breadcrumbs
1 tablespoon chopped parsley
0.5 kg/1 lb/1 lb pork sausages, skinned
½ teaspoon chopped fresh lemon thyme
25 g/1 oz/¼ cup walnuts, roughly chopped
grated rind of 1 lemon
salt and pepper
1 (1.5-kg/3½-lb/3½-lb) chicken, boned
10 pimento-stuffed olives
mustard and cress and chopped peppers to garnish

Sauté the onion and celery in the butter for about 10 minutes. Stir in the breadcrumbs, parsley, sausages, thyme, walnuts, lemon rind, salt and pepper. Mix well with your hand. Put this stuffing into the centre and legs of the boned chicken, pushing the olives in at intervals. Reshape the chicken and sew up. Place in a roasting tin and cook in a moderately hot oven (200°C, 400°F, Gas Mark 6) for 1¼–1½ hours.
To serve at once: Remove the chicken from the roasting tin and leave to become cold before cutting into slices. Garnish.
To freeze: Cool, wrap in foil, seal, label and freeze.
To thaw: Place in the refrigerator overnight and then leave out at room temperature for 4 hours before slicing.

Serves 8–10

Rump steak galantine

METRIC/IMPERIAL/AMERICAN
0.5 kg/1 lb/1 lb lean rump steak
225 g/8 oz/½ lb lean cooked ham
1 large onion
3 large tomatoes, peeled
50 g/2 oz/1 cup fresh brown breadcrumbs
2 eggs
1 teaspoon salt
¼ teaspoon pepper
1 tablespoon tomato purée
1 teaspoon dried mixed herbs
300 ml/½ pint/1¼ cups aspic jelly (optional)
salad vegetables to garnish

Mince the meats, onion and tomatoes, and mix with the next 6 ingredients. Stir until well blended. Place in a lightly oiled 1-kg/2-lb loaf tin and cover with 2 layers of foil. Cook in a cool oven (150°C, 300°F, Gas Mark 2) for 1¼ hours. Chill overnight, weighted down slightly.
To serve at once: Turn out and brush with aspic jelly, if liked. Arrange on a bed of lettuce with tomatoes, and garnish with sliced cucumber and radish.
To freeze: Turn out, wrap in foil, seal, label and freeze.
To thaw: Overnight in the refrigerator. Finish as above.

Serves 6–8

Salmon mousse

METRIC/IMPERIAL/AMERICAN
1 kg/2 lb/2 lb fresh or frozen salmon, thawed
salt and pepper
600 ml/1 pint/2½ cups milk
1 small onion, sliced
2 bay leaves, 6 peppercorns
100 g/4 oz/½ cup butter
100 g/4 oz/1 cup plain flour
15 g/½ oz/2 envelopes gelatine
3 tablespoons/3 tablespoons/¼ cup hot water
juice of ½ lemon
150 ml/¼ pint/⅔ cup double cream

Season the salmon, wrap in foil and bake in a moderately hot oven (200°C, 400°F, Gas Mark 6) for 10 minutes per 0.5 kg/1 lb. Flake. Gently heat the milk with the onion, bay leaves and peppercorns to just below the boil. Infuse for 20 minutes. Melt the butter, add the flour and cook for 2 minutes, stirring. Strain in the milk. Bring to the boil, stirring, and cook for 5 minutes. Cover and cool a little. Dissolve the gelatine in the hot water and stir into the sauce with the lemon juice and flaked salmon. Mash well or liquidise. Stir in the cream, season and pour into a 1.25-litre/2-pint/2½-pint ring mould. Chill overnight. Turn out. Garnish as illustrated.
To freeze: Cover with foil and freeze. When solid, turn out and pack in a rigid polythene container. Return to freezer.
To thaw: Unpack and thaw in the refrigerator overnight.

Serves 10–12

Chicken chow mein

METRIC/IMPERIAL/AMERICAN
generous litre/2 pints/5 cups chicken stock
50 g/2 oz/½ cup cornflour
1 (376-g/13¼-oz/13¼-oz) can sweet and sour sauce
1 kg/2 lb/5 cups cooked chicken meat, diced
225 g/8 oz/2 cups whole blanched almonds, fried in oil
1 bunch spring onions, washed and chopped
225 g/8 oz/2 cups button mushrooms, washed and sliced

Pour the stock into a saucepan. Blend 2 tablespoons/2 tablespoons/3 tablespoons of the stock with the cornflour to form a paste, then return this mixture to the saucepan. Stirring continuously, bring the sauce to the boil and cook until thickened. Add the remaining ingredients, mix well and heat through for 5 minutes.
To serve at once: Serve hot with cooked bean sprouts.
To freeze: Transfer to foil containers and cool quickly. Seal, label and freeze.
To thaw: Allow the unopened foil dishes to thaw for about 30 minutes at room temperature then place in a moderate oven (180°C, 350°F, Gas Mark 4) for 45–60 minutes until heated through. Serve as above.

Serves 10–12

Risotto

METRIC/IMPERIAL/AMERICAN
1 kg/2 lb/4 cups minced beef
0.5 kg/1 lb/4 cups onions, chopped
1 kg/2 lb/4½ cups long-grain rice
2 green peppers, deseeded and diced
2 teaspoons ground turmeric
1.75 litres/3 pints/7½ cups chicken stock
600 ml/1 pint/2½ cups red wine
3 (397-g/14-oz/14-oz) cans peeled tomatoes
0.5 kg/1 lb/4 cups mushrooms, washed and sliced
salt and pepper
watercress to garnish

Heat the minced beef in a very large saucepan for 5 minutes. Add the onions and cook for 15 minutes. Stir in the rice and after 5 minutes the peppers, turmeric, stock, wine and tomatoes. Cover and cook until the rice has absorbed the liquid and is just tender – about 30–40 minutes. Add the mushrooms for a further 10 minutes. Season to taste.
To serve at once: Garnish with watercress. Serve with a salad.
To freeze: Divide between baking sheets lined with foil. Place in polythene bags and freeze until solid. Remove the risotto in the foil from the baking sheet, and replace in polythene bags. Seal, label and return to the freezer.
To thaw: Unwrap and return the risotto and foil to the baking sheets. Cover with more foil and place in a moderate oven (180°C, 350°F, Gas Mark 4) for 1 hour.

Serves 30

Picnics and packed lunches

Now, instead of having to prepare packed lunches when you are busy trying to cook the breakfast and get everybody out of the house on time, you can just take the required number of portions from the freezer, knowing they will thaw during the morning. And with supplies of picnic fare all ready, the whole family can go off for the day whenever the sun shines.

Sandwich fillings for the freezer

Avoid freezing sandwiches with fillings which discolour or go limp and watery, or tough and dry, for example lettuce, cress, tomatoes, cucumber, hard-boiled eggs and bananas.

Freeze the sandwiches with the crusts on, and remove them on thawing. Always use fresh bread. Wrap quickly so they do not dry out in the atmosphere.

FILLINGS FOR CHILDREN

These are proven favourites with children:

Grated cheese and pickle
Grated cheese with a little chopped skinned tomato mixed with tomato ketchup
Mashed tuna with cottage cheese and 1 or 2 drops vinegar (illustrated)
Minced ham with sandwich spread

ADULT FILLINGS

Cottage cheese with chopped olives and gherkins (illustrated)
Sliced beef with horseradish
Sliced tongue with pickle

To freeze: Make up the sandwiches in the usual way, buttering them well to stop the filling going through. Interleave the sandwiches with freezer layer tissue and place in polythene bags. Seal, label very clearly and freeze. Fillings containing cottage cheese should not be stored for longer than 2–4 weeks.

To thaw: Take the required number of sandwiches from the freezer in the morning and wrap in foil. The sandwiches will be thawed by lunchtime but still be nice and fresh.

Cheese flan

METRIC/IMPERIAL/AMERICAN
50 g/2 oz/¼ cup lard
100 g/4 oz/1 cup plain flour
pinch salt
1 tablespoon cold water
FILLING:
2 tablespoons/2 tablespoons/3 tablespoons sweet pickle
175 g/6 oz/1½ cups Cheddar cheese, grated
2 eggs
150 ml/¼ pint/⅔ cup milk

Place an 18-cm/7-inch flan ring on a baking sheet.
   Rub the fat into the flour and add the salt and enough cold water to make a dough. Roll out and use to line the flan ring.
   Mix the pickle, cheese, eggs and milk together and pour into the prepared flan case. Bake in the centre of a moderate oven (190°C, 375°F, Gas Mark 5) for 40 minutes or until set and firm.
   Remove the flan ring and allow the flan to cool on the baking sheet.
To serve at once: Serve cold with salad or wrap individual portions in foil for a packed lunch.
To freeze: Loosen the flan from the baking sheet, cut into portions and open freeze. When solid, wrap the portions in foil, seal, label and return to the freezer.
To thaw: Remove the required number of pieces from the freezer and leave in the foil at room temperature for 1 hour.

Serves 4–6

Baps and bangerburgers

METRIC/IMPERIAL/AMERICAN
225 g/8 oz/1 cup pork and beef sausagemeat
½ teaspoon dried mixed herbs
3 teaspoons tomato ketchup
3 rashers streaky bacon, derinded and chopped
flour for coating
salt and pepper
3 tablespoons/3 tablespoons/¼ cup oil
3 baps
butter for spreading
1 tablespoon pickle (optional)

Mix the sausagemeat with the herbs, ketchup and bacon.
   Divide the mixture into 3 cakes, each about 2 cm/¾ inch thick. Coat with seasoned flour. Fry the cakes in the oil for about 6 minutes on each side.
   Remove and drain on kitchen paper. Allow to become cold.
To serve at once: Cut the baps in half, spread with butter, and pickle if liked. Place a bangerburger on each bottom half and cover with the top half. Wrap each filled bap in cling film to keep fresh.
To freeze: Wrap the bangerburgers and baps separately in foil or interleave with freezer layer tissue and pack in polythene bags. Seal, label and freeze.
To thaw: Leave, wrapped, at room temperature for 1 hour, then finish as before, wrapping in cling film to keep them fresh and easy to pack.

Serves 3

Cheese scones

METRIC/IMPERIAL/AMERICAN
100 g/4 oz/1 cup self-raising flour
salt and pepper
½ teaspoon dry mustard
40 g/1½ oz/3 tablespoons soft margarine
50 g/2 oz/½ cup Cheddar cheese, grated
1 tablespoon chopped parsley
3 tablespoons/3 tablespoons/4 tablespoons milk
extra milk for brushing

Sift the flour into a bowl with the salt, pepper and mustard. Rub in the margarine until the mixture is like fine breadcrumbs. Add the cheese and parsley. Make a well in the centre, add the milk and mix with a fork. Knead gently on a lightly floured board and roll out to a circle about 2 cm/¾ inch thick. Brush with a little milk and mark with a knife into 6 wedges.

Place on a lightly floured baking sheet and cook towards the top of a hot oven (220°C, 425°F, Gas Mark 7) for 10–15 minutes. Cool on a wire tray. Cut into the wedges.

To serve at once: Pack individual portions into polythene bags and take on a picnic to eat with a salad.
To freeze: Place in a plastic bag. Seal, label and freeze.
To thaw: Pack, still wrapped, in with the picnic fare. They will thaw in 30 minutes at room temperature.

Serves 6

Sponge and apple squares

METRIC/IMPERIAL/AMERICAN
150 g/5 oz/⅔ cup soft margarine
175 g/6 oz/¾ cup castor sugar
1 egg
grated rind of 1 lemon
225 g/8 oz/2 cups self-raising flour
0.75 kg/1½ lb/1½ lb cooking apples
75 g/3 oz/½ cup sultanas

Cream the margarine with 75 g/3 oz/6 tablespoons sugar, until light and fluffy.

Beat in the egg and lemon rind. Gradually stir in the flour. Transfer half the mixture to an oiled 18 × 28-cm/7 × 11-inch tin. Level the surface.

Peel the apples, slice thinly and spread over the mixture in the tin. Sprinkle with the sultanas and 50 g/2 oz/¼ cup sugar. Cover with the remaining creamed mixture and level the surface. Cook in the centre of a moderate oven (180°C, 350°F, Gas Mark 4) for 1 hour.

To serve at once: Sprinkle with the remaining sugar and serve with custard. Alternatively, sprinkle with sugar, leave to become cold, cut into squares and wrap in greaseproof paper to take as part of a packed lunch.
To freeze: Sprinkle with the remaining sugar. When cold, cut into squares and wrap in foil. Seal, label and freeze.
To thaw: In wrapped portions at room temperature for 1 hour.

Serves 12

Cakes, biscuits and bread

Baked goods freeze very well. Cakes can be frozen plain, and iced after thawing or they can be finished completely and then frozen packed in rigid polythene containers to give protection during storage. When freezing decorated cakes without a rigid container, remember to open freeze before wrapping closely and to unwrap while the icing is still solid.

Chocolate sandwich cake

METRIC/IMPERIAL/AMERICAN
110 g/4 oz/½ cup castor sugar
110 g/4 oz/½ cup soft margarine
110 g/4 oz/1 cup self-raising flour, sifted
1 teaspoon baking powder
1 tablespoon cocoa powder
2 large eggs
ICING:
225 g/8 oz/1¾ cups icing sugar, sifted
75 g/3 oz/6 tablespoons soft margarine
2 tablespoons/2 tablespoons/3 tablespoons hot water
1½ tablespoons/1½ tablespoons/2 tablespoons cocoa
chocolate vermicelli to decorate

Line the bases of two 18-cm/7-inch sandwich tins and brush with oil. Place all the cake ingredients in a bowl and beat well. Divide the mixture between the tins. Place in the centre of a moderate oven (160°C, 325°F, Gas Mark 3) for 25–30 minutes. Turn out and cool. Place the icing sugar and margarine in a bowl and beat in the hot water blended with cocoa.
To serve at once: Sandwich the cakes together with a third of the icing. Spread most of the remainder all over the cake and decorate as illustrated.
To freeze: Separate the layers with greaseproof paper, wrap in foil, seal, label and freeze. Freeze icing separately.
To thaw: Unwrap the cake and thaw cake and icing for 2 hours at room temperature. Finish as above.

Makes 1 (18-cm/7-inch) round sandwich cake

Raisin and ginger cake

METRIC/IMPERIAL/AMERICAN
100 g/4 oz/¾ cup ginger preserved in syrup, chopped
225 g/8 oz/2 cups self-raising flour, sifted
50 g/2 oz/⅓ cup seedless raisins
150 g/5 oz/½ cup plus 2 tablespoons butter
100 g/4 oz/½ cup soft brown sugar
2 large eggs
¼ teaspoon salt
1 teaspoon ground ginger
50 g/2 oz/⅓ cup mixed peel
4 sugar lumps
castor sugar to decorate

Line and grease a 20-cm/8-inch round cake tin. Drain the ginger on kitchen paper and toss lightly in a little flour. Add raisins.
  Cream the butter with the sugar, beat in the eggs and gradually stir in the flour, salt, ground ginger, fruit and peel. Turn into the tin. Crush the sugar lumps and sprinkle over the top. Bake in the centre of a moderate oven (160°C, 325°F, Gas Mark 3) for 1¼–1½ hours, or until a skewer inserted into the centre of the cake comes out clean. Allow to stand for 5 minutes. Turn out on to a cake rack.
To serve at once: Sprinkle with castor sugar and, if liked, tie a yellow ribbon around the cake.
To freeze: Open freeze, then wrap in foil and a polythene bag. Seal, label and return to the freezer.
To thaw: Unwrap and thaw for 4 hours at room temperature.

Makes 1 (20-cm/8-inch) round cake

Coffee walnut cake

METRIC/IMPERIAL/AMERICAN
3 eggs
75 g/3 oz/6 tablespoons castor sugar
75 g/3 oz/¾ cup plain flour
¼ teaspoon salt
2 tablespoons/2 tablespoons/3 tablespoons apricot jam
150 g/5 oz/½ cup plus 2 tablespoons unsalted butter
100 g/4 oz/scant cup icing sugar, sifted
1 tablespoon instant coffee dissolved in 3 teaspoons boiling water
75 g/3 oz/¾ cup shelled walnuts

Line and grease two 20-cm/8-inch sandwich tins. Whisk the eggs and sugar in a mixing bowl over hot water until the whisk leaves a trail. Using a metal spoon, fold in the flour and salt. Divide the mixture between the tins. Bake in a hot oven (220°C, 425°F, Gas Mark 7) for 10–15 minutes. Cool.
To serve at once: Spread one sponge with apricot jam. Cream the butter and icing sugar, then beat in the coffee. Spread a little cream over the jam and sandwich the cakes. Coat the cake with most of the cream. Halve 8 walnuts and reserve. Chop the remainder and press on to the sides. Mark a pattern on the top. Decorate as illustrated.
To freeze: Place on foil instead of cake stand. Finish as above except for sprinkling the icing sugar. Open freeze, then pack in a rigid polythene container. Seal, label and return to the freezer.
To thaw: Unpacked for 4 hours. Sprinkle with icing sugar.

Makes 1 (20-cm/8-inch) round cake

Gingerbread

METRIC/IMPERIAL/AMERICAN
100 g/4 oz/½ cup margarine
75 g/3 oz/6 tablespoons soft brown sugar
175 g/6 oz/½ cup black treacle
100 g/4 oz/⅓ cup golden syrup
1 teaspoon ground ginger
½ teaspoon ground cloves
½ teaspoon ground nutmeg
100 ml/4 fl oz/½ cup milk
2 eggs
225 g/8 oz/2 cups plain flour, sifted
3 tablespoons/3 tablespoons/¼ cup ginger marmalade
½ teaspoon bicarbonate of soda
1 tablespoon hot water
crystallised ginger to decorate

Line and grease a 20-cm/8-inch square cake tin. Heat the margarine with the sugar, treacle, syrup and spices. Add milk.

Beat the eggs together in a large bowl and add the warmed ingredients, flour, marmalade and bicarbonate of soda dissolved in the hot water. Pour into the cake tin and bake in the centre of a moderate oven (160°C, 325°F, Gas Mark 3) for 1½ hours. Cool. Cut into squares and decorate with slices of ginger.

To freeze: Wrap the cold uncut cake in foil and then place in a polythene bag. Seal, label and freeze.

To thaw: Leave wrapped only in foil for 3–4 hours at room temperature. Unwrap and finish as above.

Makes 1 (20-cm/8-inch) square cake

Cherry loaf

METRIC/IMPERIAL/AMERICAN
225 g/8 oz/2 cups glacé cherries
225 g/8 oz/2 cups self-raising flour
175 g/6 oz/¾ cup soft margarine
175 g/6 oz/¾ cup castor sugar
3 standard eggs
¼ teaspoon vanilla essence
6 sugar lumps, crushed

Wash and dry the cherries and toss in a little of the flour. Line and grease a 1-kg/2-lb loaf tin.

Beat together the margarine, sugar, flour, eggs and vanilla essence for 2 minutes. Add the cherries.

Put the mixture into the prepared tin and level the surface. Scatter the crushed sugar over the top and bake in the centre of a moderate oven (160°C, 325°F, Gas Mark 3) for 1¾ hours. The loaf is cooked when a skewer inserted into the centre comes out clean. Turn out on to a wire rack to cool, then remove the paper.

To serve at once: Cut into the required number of slices.

To freeze: When the loaf is cold, wrap in foil and put in a polythene bag. Seal, label and freeze. If liked, the loaf could be cut into slices and frozen with each slice individually wrapped in foil.

To thaw: Leave wrapped in the foil only for about 4 hours at room temperature.

Makes 1 (1-kg/2-lb) loaf

Date and walnut loaf

METRIC/IMPERIAL/AMERICAN
50 g/2 oz/3 tablespoons golden syrup
25 g/1 oz/2 tablespoons butter
50 g/2 oz/¼ cup castor sugar
150 g/5 oz/1¼ cups self-raising flour
¼ teaspoon mixed spice
100 g/4 oz/⅔ cup dates, chopped
50 g/2 oz/½ cup walnuts, chopped
1 large egg
2 tablespoons/2 tablespoons/3 tablespoons milk
butter for spreading

Line and grease a 0.5-kg/1-lb loaf tin. Gently heat the golden syrup with the butter and sugar until the butter is dissolved. Remove from the heat.

Sift the flour into a bowl with the spice. Stir in the dates, walnuts, the syrup mixture and the egg beaten with the milk. Mix well. Pour into the tin and cook in the centre of a moderate oven (160°C, 325°F, Gas Mark 3) for 45 minutes. Turn out on to a wire rack. Remove the paper and cool.
To serve at once: Serve in slices spread with butter.
To freeze: When cold, wrap in foil. Seal, label and freeze.
To thaw: Leave wrapped at room temperature for 3 hours.

Makes 1 (0.5-kg/1-lb) loaf

Wholemeal scones

METRIC/IMPERIAL/AMERICAN
50 g/2 oz/¼ cup soft margarine
100 g/4 oz/1 cup plain white flour
100 g/4 oz/1 cup plain wholemeal flour
2 teaspoons baking powder
pinch salt
5 tablespoons/5 tablespoons/6 tablespoons milk
1 standard egg
milk to glaze
blackcurrant or apricot jam and whipped cream to decorate

Place all the ingredients in a mixing bowl and mix together with a wooden spoon.

Turn on to a lightly floured board and knead gently until smooth. Roll out to a thickness of 1 cm/½ inch and cut into 5-cm/2-inch rounds. Place on a lightly floured baking sheet. Brush the tops with a little milk and bake towards the top of a hot oven (220°C, 425°F, Gas Mark 7) for 12–15 minutes. Turn on to a wire rack to cool.
To serve at once: Split in half and spread with blackcurrant or apricot jam topped with whipped cream.
To freeze: Pack the cooled scones in a polythene bag. Seal, label and freeze.
To thaw: Remove the required number from the freezer and leave to thaw at room temperature for 30 minutes–1 hour. Finish as above.

Makes 24–28 halves

Tuilles

METRIC/IMPERIAL/AMERICAN
50 g/2 oz/½ cup almonds, blanched or flaked
65 g/2½ oz/5 tablespoons butter
65 g/2½ oz/5 tablespoons castor sugar
40 g/1½ oz/6 tablespoons plain flour

Chop the almonds roughly. Cream the butter and sugar together until light and fluffy, add the flour and almonds and mix well. Place small heaps of the mixture, about the size of a walnut, well apart on 2 greased baking sheets. Flatten each with a wet fork.

Bake one sheet at a time in the centre of a moderately hot oven (200°C, 400°F, Gas Mark 6) for about 5-8 minutes. Leave to cool for a few seconds before removing with a palette knife and gently curve round a rolling pin. Leave until cool.

Cook the second batch and continue in this way until all the mixture has been used up.

To serve at once: Serve immediately or keep for a short time in an airtight container.
To freeze: Place carefully in a rigid polythene container. Seal, label and freeze.
To thaw: Unpack and arrange on a plate just before serving.

Makes about 20 biscuits

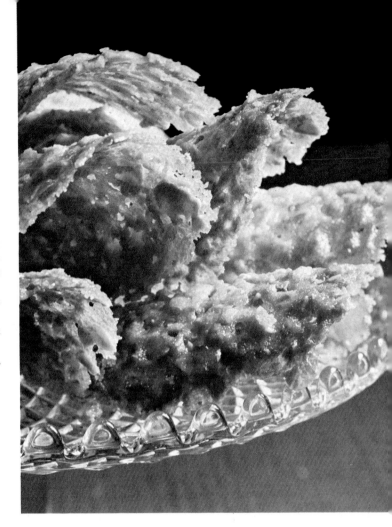

Chocolate and vanilla pinwheels

METRIC/IMPERIAL/AMERICAN
110 g/4 oz/½ cup castor sugar
110 g/4 oz/½ cup butter, softened
1 large egg
225 g/8 oz/2 cups self-raising flour
1 tablespoon cocoa powder, sifted
½ teaspoon vanilla essence
1 egg white

Mix the sugar into the softened butter and beat in the egg. Stir in the flour, knead lightly and divide in half. Knead the cocoa into one-half and the vanilla into the other. Roll out the chocolate dough between 2 sheets of lightly floured greaseproof paper to a rectangle 20 × 28 cm/8 × 11 inches. Repeat with the vanilla dough. Remove both top sheets of greaseproof and brush the chocolate dough with egg white. Place the vanilla dough on the chocolate, remove the greaseproof on top and roll the doughs like a Swiss roll, using the remaining greaseproof to aid rolling. Wrap in foil and chill.

To serve at once: Cut into 5-mm/¼-inch slices. Bake for 10-15 minutes on a greased baking sheet in a moderately hot oven (190°C, 375°F, Gas Mark 5). Cool on a wire rack.
To freeze: Freeze the uncooked dough wrapped in foil or pack the cold cooked pinwheels in a rigid polythene container.
To thaw: Leave wrapped at room temperature for 3-4 hours.

Makes about 34 biscuits

White bread

METRIC/IMPERIAL/AMERICAN
50 g/2 oz/¼ cup margarine or lard
0.75 kg/1½ lb/6 cups strong white flour
2 teaspoons salt
1 teaspoon sugar
15 g/½ oz/½ cake fresh yeast
450 ml/¾ pint/2 cups lukewarm water

Rub the margarine or lard into the flour and salt. Add the sugar and then the yeast blended with the water. Mix to a soft dough. Knead well for 10 minutes, then place in a lightly oiled polythene bag. Tie loosely. Leave to rise until doubled in size – 40–45 minutes in a warm place or about 2 hours at room temperature. Knead well for 2 minutes.

Using half the dough, roll out into a straight piece and tie loosely in a single knot. Place on a greased and floured baking tray in an oiled polythene bag and leave to rise for 30–40 minutes in a warm place. Remove from the bag and bake in a hot oven (220°C, 425°F, Gas Mark 7) for 25 minutes.

Divide the remaining dough into 36 balls and press them together in threes on a greased and floured baking sheet. Cover with oiled polythene and leave until doubled in size. Remove polythene and bake as above for 15–20 minutes.
To freeze: Wrap the cold loaf in foil. Open freeze the rolls, then pack in a polythene bag. Seal, label and freeze.
To thaw: Defrost the loaf, unwrapped, at room temperature for about 4 hours; the rolls for 1½ hours.

Makes 1 (0.5-kg/1-lb) loaf and 12 rolls

Brown bread

METRIC/IMPERIAL/AMERICAN
25 g/1 oz/2 tablespoons lard or margarine
0.75 kg/1½ lb/6 cups plain wholemeal flour
225 g/8 oz/2 cups strong white flour
2 teaspoons salt
2 teaspoons sugar
2 teaspoons black treacle
25 g/1 oz/1 cake fresh yeast
600 ml/1 pint/2½ cups warm water

Rub the lard or margarine into the flours and salt.

Add the sugar, the treacle and the yeast blended with the water. Mix to make a soft dough. Knead well for 10 minutes on a lightly floured surface. Divide the dough into two-thirds and one-third. Arrange the larger piece in a greased 18-cm/7-inch round cake tin and the smaller piece in a greased 0.5-kg/1-lb loaf tin.

Place the tins in lightly oiled polythene bags. Tie loosely. Leave to rise until doubled in size – about 40–45 minutes in a warm place or 1½–2 hours at room temperature. Remove the bags and dust the bread with flour, if liked. Place in a hot oven (220°C, 425°F, Gas Mark 7) and bake the large loaf for 35 minutes and the small one for 25 minutes, or until they shrink slightly in the tins. Turn on to a wire rack to cool.
To serve at once: Store in a bread container until needed.
To freeze: When cold, pack in a large polythene bag.
To thaw: Leave unwrapped at room temperature for 4 hours.

Makes 1 (1-kg/2-lb) round loaf and 1 (0.5-kg/1-lb) loaf